Preparing Science Teachers
Through Practice-Based
Teacher Education

Preparing Science Teachers Through Practice-Based Teacher Education

Edited by
DAVID STROUPE
KAREN HAMMERNESS
SCOTT MCDONALD

HARVARD EDUCATION PRESS
CAMBRIDGE, MA

Paperback ISBN 978-1-68253-530-1
Library Edition ISBN 978-1-68253-531-8

Library of Congress Cataloging-in-Publication Data is on file.

Published by Harvard Education Press,
an imprint of the Harvard Education Publishing Group

Harvard Education Press
8 Story Street
Cambridge, MA 02138

Cover Design: Endpaper Studio

Cover Image: FatCamera/iStock via Getty Images

The typefaces in this book are Adobe Garamond Pro and Myriad Pro.

Core Practices in Education

Series edited by Pam Grossman

OTHER BOOKS IN THIS SERIES

Teaching Core Practices
in Teacher Education
Edited by Pam Grossman

Contents

Practice-Based Science Teacher Education

DAVID STROUPE, KAREN HAMMERNESS, AND SCOTT MCDONALD

During a recent methods class in a university-based teacher preparation program, a team of four preservice teachers was leading an investigation about the effects of weathering on rocks. These team members were the lead instructors during an extended pedagogical rehearsal in which their peers—fellow preservice teachers in the same methods class—participated as students during the lessons. Prior to the current rehearsal, the preservice teaching team facilitated a lesson in which students shook jars containing rocks and different liquids to investigate how rock structure changes because of physical and chemical changes.

Unbeknownst to the preservice teacher instructional team, one group of students had hidden a jar with water and rocks on a windowsill over the weekend, hoping to observe if any changes occurred to the rock given longer exposure to water. This secret action became meaningful when the instructional team began a whole-class discussion about rock sediments during the weathering lesson. Suddenly, a student interrupted the conversation to say, "Oh wait, is our cup still in here?" Immediately, multiple students echoed her question: "Oh yeah, is that jar around?" "Oh, right, is the jar here?" The preservice teacher instructional team, however, said nothing, as students scurried to retrieve the jar from the windowsill. A student placed the jar on a desk, and six other students gathered around to look, all stating observations and hypotheses about the rock. The instructional team, one of whom held an uncapped marker in her hand because she was writing down student thinking on a poster board, remained frozen at the front of the classroom while students excitedly shared ideas. Clearly, this was an unexpected moment of student talk and action, and the instructional team seemed stuck on how to proceed.

After one minute of overlapping talk, a student finally asked an instructor: "Do you have a picture of the other jars from last week?" At this question, the marker-holding instructor sprang into action, saying, "Yes, let me find one." While this instructor found the photo on her phone, the students told their peers about the purpose of leaving the jar on the windowsill. In addition, another member of the instructional team arrived at the table and asked a series of questions aimed at pushing the students to think more deeply about their observations. After a round of questioning, a third member of the instructional team yelled out, "Okay, let's bring it back," and all eyes returned to the front of the room and the poster board—the descriptions of student thinking still in midsentence.

After this lesson ended, the students had an opportunity to debrief with the instructional team about their pedagogical decisions, and as expected, questions quickly focused on the "hidden jar" episode. One question seemed especially important; a student asked, "Why did you find a picture of the jar rather than stop that talk?" The instructor paused, then replied, "I realized that I could help, and that it gave me a way to link a clear but unexpected student interest to the lesson we were trying to teach." Then, turning to the teacher educators, that same instructor asked: "How do I do that better? How do I help students when they say something I didn't expect? How do I show them that I want their ideas and their thinking?"

We present the "hidden jar" vignette to begin this book about core practices and practice-based science teacher preparation because the story illustrates the difficulty and importance of preparing preservice teachers to foster classrooms in which students feel safe, valued, and encouraged to take actions to advance their thinking. Two themes that emerge from the "hidden jar" story provide the foundation for this book. First, the preservice teachers in this story identify important problems and opportunities of practice, and ask questions that reflect new science teaching and learning expectations (including policy documents such as the *K–12 Framework for Science Education* and the *Next Generation Science Standards*).[1] Rather than framing teaching as controlling students' thinking (e.g., fixing "misconceptions") or regulating their every move (e.g., focusing on classroom management), the preservice teachers were learning core instructional practices aimed at building and sustaining collaborative learning communities in which students' ideas and talk are the driving force of teaching. Second, the "hidden jar" story embodies an opportunity designed by teacher educators for the preservice teachers to rehearse complex core practices in a setting that simultaneously is safe and presses the preservice teachers to think differently about teaching and learning. The teacher educators structured the methods class so that preservice teachers learned to value and attend to student thinking, to recognize and make principled decisions when encountering uncertainty, and to feel comfortable asking colleagues for help. In other words, the "hidden jar" story emerged from the teacher educators' purposeful design of learning opportunities for preservice teachers to develop a vision of equitable teaching, and to rehearse core practices in a methods course.

As a community of teacher educators, we recognize that designing such opportunities for preservice teachers is crucial, but difficult. In addition, conversations among teacher educators about their pedagogies, innovations, and learning opportunities are not often elevated to public planes of discussion. Therefore, this book embodies our aim as a community of colleagues to continually and collectively improve our work as teacher educators to better prepare preservice teachers. This book is unique because we, as teacher educators in multiple preparation programs, are living the daily reality of preparing new science teachers while engaged in conversations across institutions. Through discussions over multiple years, we arrived at shared problems and opportunities of practice around teacher education: How do we help preservice teachers learn through practice-based teacher education? How do we advance as a community of teacher educators through scholarship and inquiry into our teaching? How do we learn with and from each other about practice-based teacher preparation, and continually improve our work as teacher educators? Throughout this book, we will tell stories of our collective learning as well as consider how we can continue to improve our teaching and research.

PRACTICE-TEACHER PREPARATION

Educators are being asked to reimagine teaching to support the learning of all students by engaging them in cognitively demanding tasks that are authentic to disciplines. This teaching has been described as *ambitious* because of its attention to eliciting and supporting *all* students' thinking as the foundation of ongoing sensemaking while they participate in learning activities.[2] Note that such instruction differs from other notions of science teaching, such as 5E, inquiry or "hands on," and project-based teaching, through its emphasis on students' participation in and shaping of science practices, as well as on classrooms that use students' ideas as the foundation of the learning community.

Concurrent with the emerging understanding of teaching and learning, the field of teacher education has undergone a major shift in thinking about how preservice teachers can learn about instruction. There has been a move away from a focus on developing teacher knowledge in various forms, and toward the enactment of core teaching practices based on a growing body of research about how students learn.[3] The term *core practices* as used here draws upon the definition proposed by Windschitl and Calabrese Barton:[4]

> By *teaching practices* we mean the recurring professional work devoted to planning, enacting, and reflecting on instruction. We emphasize teaching as practice as a way to acknowledge that student participation and learning are mediated most directly by teacher decisions about the kinds of tasks, talk, and tools used in the classroom. Without some common framework to describe and guide good teaching—as practice—it is difficult for either researchers or practitioners to communicate about meaningful classroom problems, and it is especially difficult for professional knowledge to be shared, tested, and refined over time.

Focusing on core practices and practice-based teacher preparation addresses what Kennedy refers to as the "problem of enactment."[5] In brief, the problem of enactment describes the difficulties preservice teachers encounter when trying to align concepts and principles they are learning from reading about, watching, and discussing teaching with the daily work of teaching actual students. Grossman and McDonald have argued for an approach to address the problem of enactment by characterizing the ways in which teacher educators can engage preservice teachers in learning opportunities to take up ambitious practices in the classroom.[6] They suggest teacher educators employ wide repertoires of teacher education pedagogies to support preservice teachers in investigating teaching and learning, situated in artifacts of practice such as case-based learning; examining lesson plans and student work; and using video of classroom instruction. However, while these approaches can impact how new teachers talk about practice, what is less well understood are the ways teacher educators can effectively prepare preservice teachers to enact and adapt these practices. Grossman and McDonald refer to this as a move from pedagogies of investigation to pedagogies of enactment.

We propose that a practice-based approach to teacher preparation organizes the work of teaching and teacher education around core practices of K–12 instruction and uses an ensemble of teacher educator pedagogies, such as representing these practices, engaging preservice teachers in rehearsals of these practices, and coaching in clinical settings, to support the approximation of complex forms of teaching over time by novices. These core practices differ across disciplinary areas as well as research groups; however, they are potentially powerful resources for helping shape more systematic preparation of preservice teachers because they are specific characterizations of everyday aspects of teaching that maintain the complexity of classroom interactions by allowing for professional judgment rather than demanding a strict adherence to scripts or routines. Importantly, teacher educators are also trying to help preservice teachers develop conceptual frameworks for teaching, which provide a rationale for and an intellectual understanding of practice. Thus, practices are represented, decomposed, and approximated, and preservice teachers engage in deep intellectual work to develop a conceptual understanding of good teaching.[7]

For preservice teacher educators, a focus on core practices is one way to make visible a skilled core of teaching.[8] The professional work associated with ambitious teaching becomes more easily appropriated by preservice teachers when it is framed as a set of practices that are explicit, principled, and adaptable to different instructional settings. Just as important, variations of core practices allow teacher educators to spend more concentrated time representing these practices, identifying other resources linked with these practices (tools for use in the classroom, videos of enactments of these practices in classrooms, samples of student work as the outcomes of these practices, etc.), having preservice teachers rehearse these practices and get feedback, and trying out the practices in classrooms. Thus, a practice-based approach might advance teaching and teacher preparation through careful experimentation and feedback to the professional preparation community.

Teacher preparation in a sociopolitical context

While we advocate for and study practice-based teacher preparation, we recognize that teacher education exists in a sociopolitical context in which there are questions raised about the value of teacher education. Specifically, there has been a rise in a free market ideology suggesting that teacher preparation should become a marketplace in which certification is commodified. Our worry is that an approach to teacher preparation that is eager to quickly certify teachers will sacrifice opportunities for preservice teachers to make sense of educational theories and practices and to participate in pedagogical rehearsals. We recognize that a common critique of teacher preparation programs is that there is insufficient evidence to support claims about teacher success. Therefore, we propose that it is incumbent upon teacher educators and teacher education scholars to continue conducting in-depth research that helps us better understand how practice-based teacher preparation may support the goals of public education, and may further help students who are already being taught by underprepared teachers and/or are learning in schools that those with power choose to marginalize.

Inspired by the multiple conversations around teacher preparation occurring among teacher educators, policy makers, parents, teachers, and school/district administrators, this book represents the genesis of a community of colleagues aiming to articulate what we have learned about practice-based science teacher preparation, and to advocate a stance of collaborative learning as we advance together. We are a group of teacher educators who are actively teaching and examining our instruction using shared language, practices, and tools. Our goal is to tell the story of how a community of colleagues can generate, test, revise, and disseminate research about core practices and science teacher preparation. In addition, we aim to be candid about our work—what fits and starts have we encountered, what puzzles and concerns us, what we are learning, and what we want to work toward as we prepare the next generations of professionals.

SIX PRINCIPLES ANCHOR THIS BOOK

To tell our stories, this book is anchored around six principles that are reflected in our research and work with preservice and in-service teachers.

Principle 1. Teaching is not natural, not magic, nor is anyone "born to teach." Instead, teaching is a learnable profession in which preservice teachers develop knowledge, practices, identities, and agency to launch their careers.

During science teacher preparation, preservice teachers also learn to empower students to participate in, question, challenge, and reshape the cultures of science and school.

Principle 2. A key feature of classrooms in which students reshape school and science cultures is that their ideas and experiences are the primary drivers of teaching and learning opportunities. Teachers help orchestrate multiple forms of student talk and sensemaking that are often constrained by the norms and values of schools, which frequently value the recitation of information. As teacher educators, we want to provide opportunities for preservice teachers to reimagine their image of productive talk in classrooms, and to rehearse the orchestration of such talk during preparation programs.

Principle 3. As a community of teacher educators, we are engaged in purposeful and productive learning with and from each other about science teacher preparation. We embrace an inquiry stance about our teacher education pedagogies, and talk in a tone of curiosity about our work. We are never finished learning. This book represents where we have been and where we are now, and previews where we hope to go to better prepare science teachers.

Principle 4. An important reason why we can grow as a community is that we use and adapt a shared pedagogical language, practices, and tools. We have common instructional objects to work with and on, thus enabling us to produce knowledge about teaching that can be developed and shared across preparation programs. In addition, we learn with and from our current and former preservice teachers as they try out, adapt, and create practices and tools in classrooms. This cycle of preparing teachers and learning with and from them, their students, and their communities creates a group of colleagues who are engaged in collective work to improve science teaching and learning.

Principle 5. These shared practices, which we describe as core practices (see the definition in the next section), are *not* "best practices." For our community, the term *best practices* implies a false narrative about teaching—that ideal or perfect instructional practices exist that teachers can use regardless of context to guarantee all students' achievement. We argue that "best practices" are too vague, prescribed, individual (i.e., not connected to a larger framework of instruction), and perpetuate inequities. In contrast, our framing of core practices is that they are designed and adapted by teachers, grounded in research on student learning, connected to short- and long-term instructional goals, provide teachers with concrete features to work toward equity and justice in classrooms, and prioritize the building of relationships with students. The core practices described in this book provide a picture of our collective and current thinking about teaching and teacher preparation. Our stance is that the naming and reification of practices is important because the practices can then be discussed, investigated, improved, and adapted. Importantly, reification is not canonization—the core practices described here are not settled and finished. Core practices constantly evolve through collaboration and collective investigation and learning from practitioners and communities.

Principle 6. Preparing preservice teachers to work toward equity and justice in classrooms requires the development of a critical consciousness *and* core practices. Neither a critical consciousness nor core practices—if separate from each other—can help preservice teachers learn to become the teachers that students need. Learning instructional practices without developing a critical consciousness means that preservice teachers might devalue the relational and humanizing features of their daily work, and might think that their job is to uncritically move through a list of routines without considering students' lives, experiences, identities, and needs. Conversely, preservice teachers need reified and learnable examples of equity and social justice practices that they can begin to learn and rehearse in preparation programs. We should not hope or assume that preservice teachers will develop equitable instructional practices or work toward justice in classrooms without support. We should also not require that preservice teachers shoulder the sole responsibility for creating pedagogies to disrupt systemwide inequities. If left to create teaching practices individually, or faced with the task of changing the culture of a school, we worry that many preservice teachers will default to enacting instruction that they experienced as learners, which likely perpetuates the very inequities we aim to disrupt. Therefore, teacher educators must help preservice teachers develop a critical consciousness *and* learn, use, and adapt concrete core practices that can provide a foundation to enact the important vision of equitable and just teaching.

KEY TERMS IN THE BOOK

As noted, one crucial reason for our collaboration is that we share language, practices, and tools around teacher education. We define some key terms here, and throughout the book, we describe how such terms are made actionable and used with preservice teachers, mentors, and administrators of schools and universities.

Preservice teacher: Someone learning to become a teacher who is participating in a teacher preparation program and is not yet certified to teach.

Core practices: The adaptable pedagogical routines that preservice teachers learn to use in order to support students' ways of interacting with science and with each other. Core practices have six features: an instructional goal; a prototypical sequence of interaction with learners; tasks, characteristic talk, and tools; underlying principles that allow productive variations of practice; a recognizable role in a larger coherent system of instruction; and the expectation that they will be adapted through principled improvisation as they come into being during interactions. Importantly, core practices are made better as teachers and teacher educators learn with and from each other through the enactment of the practices with preservice teachers, students, teachers, administrators, and communities that schools serve.

Practice-based teacher education: A framing of teacher education that focuses on disrupting preservice teachers' images of science and teaching, learning core practices through multiple opportunities to rehearse teaching, and building relationships with students and their communities.

Pedagogies of teacher education: The learning opportunities designed by teacher educators for preservice teachers to learn how to enact core practices, how to build relationships with students and communities, and how to become a reflective teacher.

Ambitious Science Teaching (AST): A pedagogical framework that supports all students to engage in intellectually rigorous and equitable learning opportunities that are facilitated by core practices and tools. Ambitious Science Teaching represents one version of how core practices can be organized and learned by preservice teachers. Along with other versions of instruction that advance teaching and learning beyond memorization, AST offers powerful examples of how core practices can serve as the foundation for rigorous and equitable instruction.

Task, talk, tools: These refer to features of core practices that preservice teachers learn to pay attention to when planning and teaching. For the authors of this book, these terms mean:

- *Task:* Instructional episodes should include a complex and content-rich scenario and high expectations for student learning. Activities are designed in service of learning about overarching science ideas and supporting students in revising their ideas over time.
- *Talk:* Teachers and students should engage in purposeful talk with elaborating, questioning, and reorganizing of ideas as the goal; students' ideas are uncompromisingly treated as intellectual resources for everyone's learning.
- *Tools:* The unit and individual lessons should scaffold student and teacher reasoning. The tools should also help publicly represent student thinking.

Clinical experience: This refers to an opportunity for preservice teachers to spend time in a mentor teacher's classroom in which they have access to time, resources, and students in order to rehearse teaching. Clinical experience is a planned learning opportunity in many teacher preparation programs.

BOOK TRAJECTORY

The trajectory of this book aims to tell the story of our research and learning as a collaborative community of teacher educators. Importantly, the chapters represent the authors' joint research and teaching. In other words, as teacher educators, we simultaneously teach and study our methods courses and other learning opportunities for new and experienced teachers. This book provides us with an opportunity to share our collective scholarship and pedagogy with colleagues.

The foundation of our work is grounded in our perspectives on learning, so we begin by theorizing about how preservice teachers learn to teach in practice-based programs (chapter 1). Building on this foundation, we then describe AST as one instructional framework that was designed, and has been extensively tested, to help reify rigorous and equitable instruction for preservice teachers (chapter 2). However, we do not view AST as a static entity, and therefore, in chapter 3, we describe a critical area of growth around AST for our community, in which equity and social justice can be more specifically articulated in the preparation and instruction of preservice teachers.

To ground the principles and practices in specific examples, we then describe how teacher education pedagogies help preservice teachers focus on moment-to-moment interaction with students. Specifically, we begin by comparing two teacher preparation sites to examine differences in how they help secondary preservice teachers learn core practices (chapter 4). We then focus on one teacher education pedagogy involving an extended pedagogical rehearsal that occurs in one preparation program (chapter 5). We also examine the teacher education pedagogy in a methods course designed to support productive science talk in elementary classrooms (chapter 6).

After establishing how core practices and teacher education pedagogies look in both elementary and secondary courses, we then shift to examining practice-based teacher preparation from a programmatic perspective. We begin by describing how practice-based teacher preparation requires an asset framing of elementary preservice teachers with regard to supporting responsive teaching (chapter 7). Next, we describe how a teacher education institution took a program-level approach to core practices that were then specified for science (chapter 8). Continuing to examine program-level structures, we look at how practice-based teacher preparation is developed outside a traditional university setting in which Ambitious Science Teaching is addressed in multiple connected courses (chapter 9), and how secondary preservice teachers learn in a methods class embedded in a secondary school (chapter 10).

Finally, we examine relationships between practice-based teacher preparation and the contexts that support the learning of preservice teachers. Specifically, we examine the important relationship between mentor teachers and preservice teachers who learn in practice-based programs (chapter 11). We also consider how universities support and constrain efforts at practice-based teacher preparation though their structural and partnership choices (chapter 12). Finally, we describe how practice-based teacher preparation programs aim to impact larger organizational units, engaging with school districts to support core practices and organizational change around science teaching and learning (chapter 13). We conclude the book with a summary of key themes and potential avenues for future research, including how focusing on teaching practice can help preservice teachers focus deeply on student thinking and learning.

CHAPTER 1

How Do Teachers Learn to Teach Science in Ambitious and Equitable Ways?

KAREN HAMMERNESS, SCOTT MCDONALD,
KAVITA KAPADIA MATSKO, AND DAVID STROUPE

What kind of experiences does a preservice teacher need to create the kind of rigorous and equitable classroom that is driven by student ideas and sensemaking? And what do we, as teacher educators, need to know and do to prepare teachers to teach in such ambitious ways? In this chapter, we delve into principles of learning for preservice science teachers that undergird a practice-based approach to teacher education and share a cycle of pedagogical practices that support preservice teachers' learning of ambitious forms of science teaching.

ONGOING DILEMMAS IN TEACHER EDUCATION

A fundamental tension exists in teacher education between the teaching practice preservice teachers can envision, and the foundation preservice teachers need to reach their vision. For example, one author of this book, Karen, has spent years studying teachers' visions. Over the course of nearly seven years of interviews with teachers about their visions of good teaching and their aspirations, she found that preservice teachers spoke with passion and excitement about the kind of teaching they wanted to accomplish. But many also described profound disappointment as they gradually realized the difficulty of translating those visions into the daily classroom reality of working with their students. Karen saw that novice teachers—as early as their first few months of teaching—were already beginning to believe that their visions of good teaching were not possible, or worse, coming to suspect their students were not capable of the expansive and rich learning experiences they had hoped to provide.[1] One new teacher reflected that the needs of her students had caused her to reconsider her ambitious vision, worrying that it was "totally unrealistic." Another novice teacher, only a year into teaching, described a "huge disconnect" between what she imagined her teaching could be in comparison to what she and her students were doing every day in the classroom. She explained that the juxtaposition between her vision and her daily practice was "so depressing," noting, "There's so much more that I should and could be doing. My vision reflects what I think my teaching should be and how kids learn. And I'm not there. Which means I'm not doing everything I can to help kids learn." While their teacher education courses had introduced preservice teachers to "visions of possibility" by sharing foundational pedagogical concepts and important theoretical grounding, they did not provide an image of daily instruction that would enable the preservice teachers to enact those aspirations.[2] Preservice teachers had a reach, but no footing.

Teacher education has been roundly critiqued for years for providing too few opportunities for preservice teachers to learn to do the work they will be responsible for as teachers. Preservice teachers frequently have substantial opportunities to read about instruction and to study theories of teaching and learning that support those perspectives. Such resources can help preservice teachers to develop visions of the possible, but research demonstrates again and again they rarely have scaffolded opportunities to test out the practices they will enact as teachers in classrooms.[3] This disconnect between their vision and actual teaching practice makes their pedagogical footing unstable. Preservice teachers can aim for a vision of equitable, challenging, and rigorous work with children that reflects their principled commitments; however, without opportunities to learn about and rehearse teaching practices that will help them develop such a classroom, they can become disillusioned by the difficulty of enacting their vision. Such disillusionment can be especially dissonant for new teachers who learn about and become committed to a vision of equitable education. If they have few opportunities to rehearse or learn specific teaching practices that support equity, they can fail to advance learning in their classrooms and may come to feel unsuccessful; the gap, then, between vision and practice can become deeply discouraging.[4] Even worse, without practices for enacting equitable education, the best intentioned beginners might perpetuate rather than disrupt the social inequities that surround us.

This emphasis of teacher education programs on foregrounding key foundational ideas and inspirational visions of good teaching, but offering few opportunities to learn to enact such ideas and visions in daily interactions with students, has been an ongoing "problem of teacher education."[5] Mary Kennedy named this challenge the "problem of enactment," and addressing this challenge has led many teacher educators to argue for a more substantial focus on practice.[6] Deepening opportunities in practice is not a new idea. Teacher educators have long designed approaches that enable preservice teachers to study and rehearse aspects of teaching. In the 1960s, Nathanial Gage led a teaching lab in which teachers rehearsed the technical skills he viewed as critical to teaching. Even earlier, opportunities to learn teaching practice were deliberately incorporated into normal schools preparing teachers in the 1800s and have reemerged in nearly every era since.[7] In each of these efforts, the terms *practice-based* and *practice* take on different meanings, and we acknowledge the complexity and long history of this language as well as the dynamic intricacies of the meanings.[8] The scholarship in teacher education over the last decade that we build upon in this book has invigorated this conversation, especially in terms of scholarly efforts to identify, elaborate, and systematically study the kinds of teacher education pedagogies that help preservice teachers to learn principled and theoretically grounded pedagogical practices and to develop a professional and personal vision of teaching.

We view teaching as a professional practice, which means that it features evolving sets of routines, tools, and explicit (as well as implicit) norms for participation. To learn such professional practices, preservice teachers draw upon, contribute to, and are assessed and evaluated by others in the field, developing identities as a teacher.[9] Magdalene Lampert commented, "When one joins teaching as a practice, the learning of the activity and the acquisition of identity go hand in hand."[10] A key element of teaching culture consists of shared practices that can be learned. Learning teaching happens in community—discussing, testing, critiquing practices, and challenging decisions. Teaching is also *relational* and *reciprocal* work—teachers need to know how to notice, predict, and be open to and interact with students' ideas; to read and understand social interactions; to support and build upon productive talk; and to understand how individual and collective cultures can enhance learning. To enact the ambitious visions and foundational ideas preservice teachers learn about in teacher education, they must hone their ability to do that complex, relational work of teaching in the moment through rehearsals, feedback, and coaching by experienced teachers and teacher educators. Through scaffolded work on interactive practices that encourage principled improvisation, preservice teachers can learn the more interactive relational judgments required by the profession of teaching.[11]

For decades, teacher educators and policy makers have referred to the gap—sometimes even calling it a chasm—between theory and practice. But are theory and practice mutually exclusive? Social psychologists, anthropologists, and learning scientists continue to press against that dichotomy, arguing for a far more fluid understanding of the relationship between them: as Mike Rose noted, "Western Culture has tended to oppose technical skill to reflection, applied or practical pursuits to theoretical or 'pure' inquiry, the physical to the conceptual. As we get closer to students doing their work, however, we see how complicated these distinctions are: the technical gives rise to reflection, the physical and the conceptual blend, and aesthetics and ethics emerge continually from physical activity."[12] We offer a view of learning to teach that unsettles that long-accepted language, and frames learning to teach as responsive work, accomplished through the use of principled practices (as we describe in chapter 2) that are inextricably entwined with theoretical principles of learning.[13]

This approach is neither void of learning theory nor reliant upon scripts or manuals. The focus of learning to teach shifts toward opportunities for preservice teachers to test, revise, and gradually strengthen the activities, strategies, and daily work they will be engaged in as classroom teachers in relation to a vision of good teaching that is principled and theoretically grounded. We have found that a laser-sharp focus on student ideas—the experience of interacting with, building on, and supporting the sometimes surprising, in-the-moment authenticity of student thinking—is most likely to happen when learning in practice.

While situating teacher learning in practice might seem, at first glance, a natural response to challenges around a lack of opportunities for enactment, teacher educators have raised questions and challenges about learning a professional vision, the role of theory, and the importance of culture and context. If teacher educators overemphasize teaching as merely a set of technical skills, there is a danger that theory could be underemphasized, treated as distinctly separate, or even missing altogether—preservice teachers might not learn and understand the foundational rationale for their work and choices. A focus on practice without a concomitant connection to a theoretical foundation could lead teachers to implement strategies without any understanding of context, communities, or their students; they could potentially see their role as technicians who are simply implementing a set of practices, and thereby perpetuate preexisting inequities. This is deeply problematic given the critical importance for teachers to learn about and prioritize students' identities and cultures, and to center their voices, perspectives, and knowledge in the classroom. Finally, a focus on practice could lead to the reverse of the problem that previous research on vision has revealed: teachers who have a strong footing but no vision of enactment. Like a long history of teacher educators before us, we also worry about these questions. This book represents our current thinking, research, and approaches to practice-based teacher education. It also is an attempt to press, wonder, investigate, and try to make gains on these challenges and concerns. We hope that in this book you will see evidence of our efforts to take each of these concerns to heart; to listen and learn from our colleagues, teachers, and students; and to continue to puzzle through these challenges.

SOCIOCULTURAL PERSPECTIVES ON FOUR PROBLEMS OF TEACHER EDUCATION

A focus on learning about teaching as well as learning the practice of teaching rests on the assumption that teaching is a complex and intricate cultural activity that can be learned. Given this stance, we discuss how practice-based teacher education can support teacher educators in working with four widely documented problems of learning to teach: the apprenticeship of observation; the problem of enactment; the problem of complexity; and the problem of inequity.[14]

First, learning to teach requires that new teachers come to think about and understand teaching quite differently than they experienced it as students. Daniel Lortie called this problem "the apprenticeship of observation," referring to the learning that takes place by virtue of being a student for twelve or more years in traditional classroom settings.[15] Second, the "problem of enactment" recognizes that helping teachers learn to teach in ways that reflect an ambitious and rigorous vision requires learning not only about teaching, but learning the practice of teaching. Third, the "problem of complexity" recognizes that the contingent and personal nature of learning means that teachers must always be using judgment, compassion, and knowledge in the moment. We also suggest a fourth problem of learning teaching: the "problem of inequity." In numerous classrooms, nationally and internationally, the cultural assets and funds of knowledge students bring to the classroom, and the potential they have as sensemakers, are not fully realized and utilized.[16]

We have found a sociocultural perspective that foregrounds the enactment of practice as well as the development of identities is especially helpful in puzzling through these four problems. A sociocultural lens helps us draw attention to the ways that preservice teachers' learning develops through participating in a culture and community (just as pupils learn), learning practices, and building professional identities. This lens makes salient the central role of tools, resources, and other learners in shaping and supporting preservice teacher development.[17] Sociocultural perspectives draw attention to the role of identity—and to the fact that struggling to understand teaching requires deep engagement with one's own identity and ways of being. It also means committing to learn about and emphasize the identities and cultures of students, and to developing knowledge of and with the community and schools. Importantly, a sociocultural perspective means explicitly recognizing this identity work as a critical part of supporting a transition to equitable teaching practices.

For preservice teachers, developing professional teacher identities is a uniquely challenging pursuit, given their long apprenticeship of observation. Few preservice teachers have experienced teaching in the ways we encourage them to practice. For many, paying attention to student learning and student thinking is a new endeavor. Being an "expert" and "knower" is more frequently a central aspect of their initial definition of good teaching, which can undermine preservice teachers' ability to notice and interact with student ideas and prior knowledge. Reflecting on their own identities, strengths, and needs in relation to learners may also be a new experience, as noted in chapter 3, both for those who have experienced oppression and those whose worlds were fairly aligned. Yet becoming a teacher means learning ways of seeing and acting that do not come naturally, and that run counter to more typical or mainstream ideas that center good teachers as presenters of knowledge and the sole holders of expertise.

Taking a sociocultural perspective on preservice teacher learning also means we start from the assets that preservice teachers bring to their learning, rather than a more deficit view of what they lack. While we know that preservice teachers' prior knowledge of teaching is shaped by their apprenticeship of observation and is rooted in their own initial identities as students and learners, our work focuses on their agency as sensemakers. Our intent is to view them as empowered learners, many of them coming to teaching with their own visions, commitments, resources, and knowledge. The contextual and cultural nature of teacher education also foregrounds learning in *activity*: this means we design learning experiences grounded in authentic activity and practice through the use of a variety of teacher education pedagogical practices and by embedding preservice teacher learning experiences in settings with students.

We aim for preservice teachers to develop a conceptual model of teaching as a phenomenon in much the same way that we want K–12 students to do with science. Preservice teachers need a theoretical framework—an understanding of how students learn; how disciplinary work is carried out; how to develop equitable and just practices; and of how students' cultures, communities, and contexts interact with the fundamental Whiteness of the current structure of schools. Preservice teachers must investigate and rehearse teaching in relationship to those theories and principles, to build up explanations and their own understandings about what makes practices productive for students. They need opportunities to gather empirical data that allows them to support those claims for themselves and for others, such as their administrators, parents, and students.[18] In this way, preservice teachers are taking what Marilyn Cochran-Smith and Susan Lytle have termed an "inquiry stance."[19] Given the intellectual demand required to understand student thinking, developing a conceptual model is especially important for new teachers because it can function as a foundation for making meaning of this complex work. Indeed, ambitious forms of teaching science, like the Ambitious Science Teaching framework we discuss in this book, can serve as conceptual models for learning teaching, functioning as a tether between a set of key practices and foundational research on students' learning. These practices—as we work on them with preservice teachers—become the foundation for reflection, analysis, and examination. In this way, teacher education pedagogies help preservice teachers engage in a process of sensemaking—but not of phenomena of science. Rather, they are making sense of teaching and learning as phenomena.

What are preservice teachers learning about teaching science?

When we engage our preservice teachers in learning with their peers, and with students, what do we want them to develop and learn? We aim to help preservice teachers develop a vision of good science teaching; a set of core science teaching practices that help them design rigorous and equitable learning opportunities; context-specific knowledge about their students, communities, and schools; and a critically conscious identity as a science teacher.

A vision of good science teaching. Many teachers we know have a vision of the kind of teaching they aim to enact that drives their practice and is grounded in their commitments. For many teachers, a vision can be a starting point for growth. Teachers are often motivated to bridge the gap between their vision of what could be and what they are currently able to do, in order to feel more efficacious and professional in their work. Of course, for many teachers, their vision is also deeply personal. Preservice teachers come in with their own commitments, values, and reasons for teaching. However, not all initial visions are clear, as Karen found, nor are they anchored to practice. Further, sometimes preservice teachers' visions can be biased or exclusionary, impeding important shifts they need to make to reach all students.

The vision of teaching we share in this book—and with our preservice teachers—is grounded in research on how students learn science. At the heart of this vision is students' meaning making from abstract ideas and complex phenomena and an intention to make students' thinking, puzzlements, and queries the center of instruction. This vision focuses on students of all racial, ethnic, class, and gender categories to reason about phenomena, to participate in building meaning, to use the discourses of the field, and to solve real problems. A context-specific or "place-based" enactment of this vision requires understanding and prioritizing of communities, students, and schools. We believe learning about this vision can help them clarify, extend, and deepen their own visions and commitments to their work.

Ambitious Science Teaching practices. Research with preservice teachers has revealed again and again that having a vision of good teaching is not enough.[20]

Without an initial or beginning set of core practices that teachers can enact, and that enable them to support student learning and success, preservice teachers come to question the visions they learn in teacher education as unrealistic or impossible. Even if they've had multiple opportunities to read about or watch videos of the most inspiring and rigorous classrooms, they still may come to believe that they can't enact such a vision; or worse, that their students are not capable of that kind of learning. For example, we view the core practices of Ambitious Science Teaching (see chapter 2) as a dynamic representation of our best thinking about a potential set of initial practices. They are rooted in research and scholarship on learning, but are constantly evolving as we learn with teacher education and teacher colleagues. This reflects the history of this work as well; these practices were identified not only by drawing upon research on what practices support student learning, but also by studying what good science teachers do. Thus we continue to learn from teachers and one another and to negotiate and learn more about these practices. Working closely with preservice teachers on these practices enables us to first learn *with them* as they test out, shift, deepen, and contribute to our collective understanding, and then to learn *from them*, as graduates. The creative negotiation and work they do in the classrooms as agentic teachers, in turn, feeds our growing collective scholarship.

Context-specific knowledge about students, communities, and schools. In all our classrooms, preservice teachers are preparing to teach in specific settings— in particular schools, in unique neighborhoods and communities and districts with rich histories, and with singular historical, cultural, and geographic assets and resources. As you will read in chapter 3, preservice teachers are learning how to prioritize students' communities, languages, and cultures by leveraging local phenomena, sometimes phenomena with political and social implications, that can serve as the material for anchoring instruction. They are seeing how an inquiry stance can help them learn about their own settings, in turn valuing and centering their students, communities, and schools. Importantly, they are learning to help students critique, challenge, and change the culture of science to create more inclusive and socially just learning experiences that scaffold participation of all students.[21] For some teachers, this kind of personal learning may be at the center of their own visions, while for others it may be quite nascent.

Identities and critical consciousness as science teachers. Through participation in the learning experiences we describe, teachers are also developing a definition of themselves as a critically conscious science teacher. This requires interrogating their own identities as well as their experiences of learning and how their conceptions of race, culture, language, socioeconomic status, gender, and physical ability shape both their science instruction and their expectations of students. Developing a critical consciousness through studies of privilege, power, and problematic dichotomies, in turn, enhances learning how to support, honor, and reinforces youth identity.[22] Conceptualizing themselves as learners— as interrogating their own identities, strengths, and needs—requires the kind of inquiry stance we emphasize. We believe that by modeling ourselves in our classes, being metacognitive, questioning and puzzling through the intricacies of classroom teaching moments, and acknowledging and examining our own positions and identities enables a setting that supports a stance of teachers as inquirers, as opposed to a vision of experts who are settled and unshakeable in their content and practice.

Where are prospective science teachers learning?

Although our preservice teachers are learning a set of shared theoretical practices and principles, they are also learning in settings that vary in histories, geographies, and cultures. We argue that strong teacher preparation takes *advantage* of the unique features of the context, the individuals, and the communities (see figure 1.1) and can be strengthened when visions across these contexts are collectively supported and when teachers learn to benefit from the specific features of the setting as well.

For instance, the articulation of the Framework for K–12 Science Education and the development of the Next Generation Science Standards is a particularly important feature of the federal and state educational policy context in the United States. The relative consistency of these foundational framing documents —grounded in research about how students from diverse backgrounds learn science and the conditions under which they can participate in both sensemaking and knowledge-building activities in the discipline—contributes to a shared vision of science teaching.[23] They call for a very different set of roles and activities for teachers, one consistent with the evolving vision of teaching and classrooms we are aiming to develop with our preservice teachers.

The potential to take advantage of specific places—to develop place-based curriculum—is critical for preservice teachers to not only understand but also

FIGURE 1.1   Contexts for preservice teachers' learning

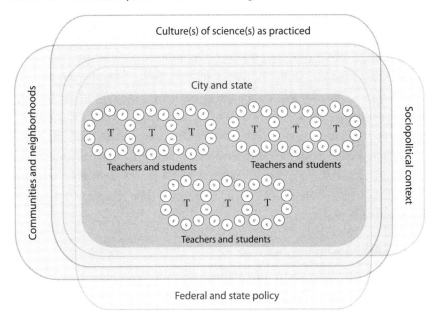

to have experience learning how to enact in practice. Teacher educators can help preservice teachers learn how to draw upon the resources and assets of local *communities and neighborhoods, or of cities and states.* For instance, in chapter 3, Kirsten Mawyer explains how she engaged preservice teachers in mapping the *ahupuaʻa,* a land division extending from a mountain to the ocean to help them understand the connection between the land and the Kanaka ʻOiwi culture. Pedagogies like community and neighborhood asset mapping enable preservice teachers to understand and surface local assets and resources. These connections between science content and place help teachers contextualize science learning in authentic ways and allow for the prioritizing and highlighting of local geographic, historical, and cultural assets.

At the innermost layer, we take advantage of and build upon *teachers' own classrooms and schools.* Teacher candidates are not isolated individuals learning on their own, but rather, they learn in community with their students, their schools, their colleagues. Yet to fully capitalize on this potential, we must take into account the considerable differences in visions and practices that can emerge across schools and universities. Adeptly avoiding the "two-worlds pitfall," as described by Sharon Feiman-Nemser and Margaret Buchmann, requires that mentors, principals, and other school faculty have clear roles in the enterprise of preparing teachers and are recognized as key participants in the development of a shared vision of teaching and learning science.[24] Yet differences in status, expertise, and roles for university teacher educators and school-based faculty can make it difficult to conceptualize all stakeholders as teacher educators, and can impede efforts to develop a shared vision and shared responsibility.[25] These include viewing variations in expertise as resources, as detailed in chapter 11 (building upon mentors' knowledge about their students and community), rather than obstacles; recognizing and working on status and power differentials; and, as in chapters 10 and 13, developing trust and relationships over time (chapter 10 details a fifteen-year relationship).

How are preservice teachers learning?

We—the teacher educators in this volume—have actively sought out teacher education pedagogies we can engage in that are theoretically grounded, are developmentally coherent, and can guide our work together. We call them "pedagogies" to avoid confusion with K–12 teaching practices. These pedagogies make up a set of learning experiences that teacher educators use with preservice teachers in the same way that preservice teachers might use science teaching practices as a system to support the learning of K–12 students. The pedagogies use ambitious forms of science teaching practices as content—they focus on inquiry into the work of teaching. Like Ambitious Science Teaching practices for science classrooms, the teacher education pedagogies are drawn from research on learning and they build on one another to influence skills, values, habits of mind, professional vision, and identity. Pedagogies are also similar to teaching practices in that they are thought of as jointly constructed with preservice teachers (the learners in this case) and mediated by tools, shared language, and conceptions of competence.[26] As with Ambitious Science Teaching practices, the teacher educator uses judgment about the pedagogies based on their program context, time, and human capacity to manage the work needed. The pedagogies, like teaching practices in K–12 classrooms, require disciplined improvisation as they are planned and enacted.[27]

Pedagogical cycles to support learning teaching (see figure 1.2) take time, and with teacher education courses typically having fewer than fifty contact

FIGURE 1.2   Cycle of pedagogies

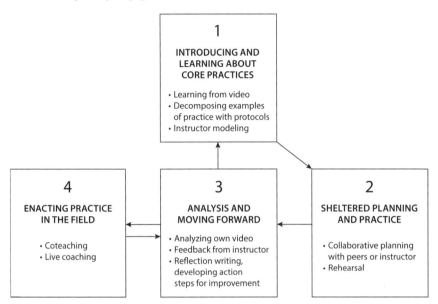

Adapted from Morva McDonald, Sarah Kavanaugh, and Elham Kazemi, "Core Practices and Pedagogies of Teacher Education: A Call for a Common Language and Collective Activity," *Journal of Teacher Education* 64, no. 5 (2013), doi:10.1177/0022487113493807.

hours in a semester/quarter, we are time constrained. We have to be selective about which practices will be the focus of study; we have to consider which are most challenging to learn, are used most frequently, and have the most impact on equitable learning opportunities for K–12 students. We use the term *core practices* to reflect our decision to focus on practices that are central to the work of teaching, and that preservice teachers might begin to learn.[28] Most of us focus on practices or aspects of practices that are responsive to students' ideas, that foster students' participation in authentic disciplinary work (investigations, data analysis, evidence-based explanations, etc.), that involve sensemaking in small groups as well as collective sensemaking as a whole class. These pedagogical cycles are introduced after we spend time eliciting preservice teachers' ideas and conceptions of teaching, and begin explorations of identity; important grounding practices are described more fully beginning in chapter 3. Just as we hope our preservice teachers will focus on eliciting student thinking, we do the same with our preservice teachers.

The purpose of these pedagogies is never to have preservice teachers mechanically reproduce specific teaching strategies. The first recommendation we give our preservice teachers for their clinical experiences is to build relationships with students. We tell them: "Don't try out anything in the front of the room until you've done the backstage work of getting to know students as both 'whole persons' and as learners." We want them to develop competence with and agency about things that matter: understanding what resources children bring to the classroom, getting learners to engage in respectful and productive talk with one another, representing students' thinking publicly, connecting to and building upon local community knowledge and resources in curriculum design. We intend for preservice teachers to gradually *pay less attention to practice itself* and *focus on responding to students' ideas, interactions, and needs.*

Figure 1.2 represents the grouping of related pedagogies into sets and how these constitute a cycle of increasingly situated and authentic experiences.[29] Each of the four sets of pedagogies provides different experiences related to teaching and learning that build on the others. Each pedagogy foregrounds different representations of teaching practice. Each amplifies parts of practice in different ways, and provides opportunities for learning. We build in supports for preservice teachers as they move from pedagogies of analyses to design, to enactments, and finally to action-oriented next steps.

Overview of the Cycle of Pedagogies

1. *Introducing and learning about core practice(s)* requires some telling, some showing, and some intensive role-playing. This often begins by reading about a practice, then examining video of instruction, preferably shot in culturally and linguistically diverse classrooms or voluntary learning environments like museums or science centers. Alternatively, transcripts related to the enactment of core practices can be analyzed. In addition to focusing on the interactions among teachers and students to accomplish particular goals, we take a critical lens to both the video and transcripts. Who is being served by this curriculum, this form of instruction? Do students have a chance to develop agency? What does this practice prepare students for? Who is on the margins of this activity? What scientific sensemaking is going on? Who is participating in that sensemaking? We always use positive case examples with teacher video, but we want our novices to know that

these questions should be asked of any classroom episodes as part of an inquiry stance on teaching.

The most dynamic pedagogy in this set is the instructor modeling the practice(s) while the preservice teachers take on the role of K–12 learners. Preservice teachers get a chance to "feel" how the intellectual work to be done is framed by the instructor and then interact with peers as they navigate the lesson together. We stop to let our preservice teachers know which moments we're puzzling about, and then open opportunities to consider the students in front of them and the context of the classroom, discuss what to do next, and consider options together as a class. This metacognitive talk opens up the black box of pedagogical reasoning so preservice teachers can hear about and engage in difficult choices they may not have been aware of and link action with underlying principles. This pedagogy is also about vulnerability. The openness of the instructor to self-critique and to questions can later encourage preservice teachers to take similar risks necessary for experimenting with responsive instruction.

2. *Sheltered planning and practice* in the form of rehearsals allows novices to shift into a lesson design mode and wrestle with choices about who will do what, when, why, with what supports, and for what purpose.[30] From our experience, preservice teachers invest great effort in planning because they know they will also enact this lesson (or parts of a lesson) with their peers acting as students. During this rehearsal, the instructor or those in the role of students can "press the pause button" to ask the "teacher" to comment about some move they've just made and perhaps to do a replay using a different strategy. Because preservice teachers will also take on the role of students, as in the "hidden jar" story (in the introduction), they also learn about core practices from a pupil's perspective as their peers teach.

3. *Analysis and moving forward* requires the novice to step back from the raw experiences of rehearsals and ask: What really happened? Why did it unfold that way? We video everyone's lesson and structure their examination of it by focusing them on their students: How did they respond to your framing or scaffolds? Who was well served—or not—because of your choices? In addition, we ask them to apply these lessons to future attempts: What are your next steps for improvement? Can you generalize what you've learned to other situations? Feedback from instructors about the teaching and about

the reflection itself deepens their learning. As part of this process, preservice teachers should collect data from students, not just about what they learned, but how they were or were not supported in learning. For rehearsals or enactments in the field preservice teachers can, for example, devise and distribute exit tickets to students. The data can then help them analyze how the practice worked, for whom, and under which conditions.

4. *Enacting practice in the field* may occur in different places in the cycle. In some programs, preservice teachers try out a practice in classrooms soon after their rehearsals and analyses. In other programs, this enactment step has to wait until they are in their field placements, perhaps weeks later. Instructors have to decide: How much should we deconstruct a complex set of practices for this enactment? With what supports? In classrooms, the teaching opportunity may focus on only one or two practices, perhaps framing an activity for students and pressing their thinking as they work in small groups. In other cases, the novice might manage a whole-class discussion and represent students' ideas publicly. However, even this structured approach requires that preservice teachers attend to real students' needs, weighing the ambiguities of what to say or do next while being responsive to students' ideas, and to put together elements of instruction that had previously been tried out separately. With the central focus on student learning and student meaning making, this approach enables, supports, and makes explicit the understanding of student identities, student communities, and student ideas and can help surface challenges and approaches that attend carefully to them. As with the rehearsals, these attempts at teaching are recorded and analyzed in specific ways, with teacher educators providing targeted feedback.

As you can see, some pedagogies in this cycle are designed to slow down or unpack acts of teaching, even replaying them with the intention of enhancing preservice teachers' ability to notice. Crucial but often overlooked aspects, such as how one interacts with children, can become objects of inquiry and improvement (how student comments are elevated or dismissed, how to respond to a puzzling comment by a student, how to connect to and emphasize local or cultural knowledge, how to draw upon students' languages). All parts of the learning cycle deprivatize teaching and open it up to critique as part of a professional ethos that preservice teachers can carry into their careers.

CHAPTER 2

The Role of Core Practices in Science Teacher Preparation

MARK WINDSCHITL, JESSICA THOMPSON,
MELISSA BRAATEN, AND DAVID STROUPE

Teachers have a singular impact on students' learning and their capacity to continue learning throughout life. Teachers, in fact, influence children's academic futures more profoundly than the kind of school a child attends, the quality of curricula, alignment of instruction to standards, class size, and every other factor associated with the educational infrastructure.[1] Much of what great teachers do is not about instruction per se, but rather about the complex and layered dynamics of working with diverse human beings who bring their whole selves to school. This requires building trustful relationships with young people whose life experiences and cultural backgrounds can be a world apart from their own, maintaining emotionally safe classroom environments for talk or just being themselves, openly sharing with students the joys of learning from them, and working minor miracles to find specialized help for children who need it.

AN EXPANSIVE VIEW OF STUDENT LEARNING

All these efforts lay the relational groundwork for effective teaching, but what about instruction itself? Accomplished educators use a range of adaptable practices to routinely foster *expansive and equitable forms of sensemaking*. For example, they start off instruction by making visible what students already understand about the science, discovering how learners' everyday experiences might relate to the topic at hand, and foregrounding their puzzlements as starting places for deep dives into the content. They help students leverage what they know, and want to know, in order to cogenerate increasingly sophisticated insights about the natural world. But there are other kinds of sensemaking that are equally critical for learning. As students begin constructing explanations for events and solving problems together, the teacher can set up spaces for conversations that give them access to their peers' reasoning. This allows them to hear how classmates think about new information or evidence and to assess the credibility of their own theories in comparison with those around them. Understanding the working minds of others is part of authentic science, in which alternative ways of viewing phenomena can add nuance to or challenge the community's current knowledge. In yet another form of sensemaking, students can figure out how the different epistemic tools of science—that is, designing investigations, analyzing data, revising models of events and processes—can work together to test ideas that matter to them. And finally, in supportive classrooms, students begin to make sense of themselves as learners. From kindergarten to high school, where the kinds of work described above are commonplace, children feel increasingly confident that they have the habits of mind, social skills, and voice to struggle productively about problems they feel are consequential.

Teaching practices in science, then, are not just about positioning students to "know stuff," but also about helping them understand how to learn in collaboration with others, to deal with the uncertainty that is endemic to science, and to question whose interests are served by the methods and outcomes of the discipline. In other words, the kinds of teaching we envision help cultivate students' identities as capable knowers who can take action to achieve challenging goals. This book is premised on the idea that *teaching toward this critical and expansive view of sensemaking is learnable.*

In this chapter we share how a pedagogical framework, referred to as Ambitious Science Teaching (AST), was developed to nurture meaningful learning for students and allow preservice educators to reimagine common ways of doing business in classrooms. It was synthesized from the literatures on teaching expertise, student learning, equity, and disciplinary activity in science. We drew strategically from these bodies of research to identify classroom conditions and opportunities for engagement that could meet the needs of increasingly diverse and underserved student populations, in both rural and urban settings. This led us to identify an initial set of practices that could shape the environment and the opportunities for learning, while also being humanly possible for educators who often see more than 130 students per day. This framework is one of several approaches represented in this book that articulate a vision of teaching and a principled set of teaching practices that work together to influence student learning. We focus on Ambitious Science Teaching as a well-documented case of core practices by providing a history of how it was developed and how it continues to evolve in order to meet the needs of students in varied classroom contexts.

From the perspective of a professional community, sharing a clear conception of what counts as a teaching practice has advantages in cases where members intend to argue about them, compare effective variations of them in new contexts, and improve them over time. The different core practices used by teacher educators in science as well as mathematics, social studies, and English language arts have similar features. Each is represented as a prototypical sequence of interactions between teachers and students, guided by both predefined and emergent goals. A core practice has characteristic combinations of tasks, talk, and tools that work together to encourage learning and participation from all students.[2] As teachers become familiar with coordinating these dimensions of practice over time, they facilitate dialogue and activity with increasing skill, and open up more accessible opportunities for learning to more students.

We want to make clear that *all* preservice teachers learn to take up practices, but this often happens by default. When preparation programs, for example, place little or no emphasis on what highly accomplished and responsive educators do in interaction with students, their novices are more likely to emulate the familiar routines of their mentors in host classrooms, and in many cases, reproduce instruction that is less attentive to equity and less supportive of genuine sensemaking than our children deserve (e.g., discourse limited to offering up right answers, students completing confirmatory labs or filling out worksheets).[3] Preservice teachers certainly need more than instructional practices to disrupt these norms, but a socially just and expansive pedagogical repertoire is one critical tool to resist and transform potentially oppressive systems of education for students and their families.

In well-designed programs, preservice teachers develop the capacity to do this work across the entirety of their preparation experience, not simply through courses on instruction or curriculum. Ideally, one's evolving practice and professional identity grow out of principles from courses on foundations, assessment, learning, and development, and from the ongoing struggle with peers to answer the question: What kind of teachers do we want to become? Coursework, however, is only the start. Everything that is read about and enacted at the university can be reinforced or called into question during the lengthy and intensive phase of clinical work, where preservice teachers have to make sense of professional practice within the dynamic and often instructionally conservative confines of K–12 classrooms. What they have learned and come to value in their programs can be tested in schools where patriarchy, racism, classism, and heteronormativity may affect students' lives every day. To deal with these realities, preservice teachers need strategies and the fortitude to elevate historically marginalized voices, seek connections across lines of difference, and empower students as agents of social change.

BOUNDARY CROSSINGS: WHAT PRESERVICE TEACHERS MAY (OR MAY NOT) BRING INTO CLASSROOMS

Fourteen years ago at the University of Washington, we, the authors, were working with a new group of enthusiastic, science-smart preservice teachers. We provided what we believed was a compelling variety of readings, videos, activities, and principles to push their thinking about teaching and learning. Eventually, our novices talked as though they were ready to make the leap into classrooms filled with real children, but we felt a growing sense of doubt about what they might carry with them into their clinical experiences. We wondered, Are our preservice teachers really learning if their interactions with students don't reflect the ideas from the months of coursework?

We decided to follow a cohort of secondary preservice teachers into their clinical experiences. To our dismay, we recognized a serious gap between what our preservice teachers talked about in methods and what they were now trying to do in classrooms. Many of them freestyled their own pedagogical routines or simply mimicked their mentors' teaching. They attempted to start classroom discussions, but such efforts were unstructured and their students often found the talk directionless or monopolized by a few peers. There was no framing, no scaffolding, no ways to make students' thinking visible or elevate their ideas. Our exasperated preservice teachers claimed they knew *what* they wanted to have happen, but didn't know *how* to make it happen. We were not even sure that this was accurate.

All preservice teachers face challenges, but our observations suggested that the methods preparation may have set them up for unproductive struggles and threatened their identities as competent beginning educators. We failed to help them take ideas from our conversations about broad classroom goals and seemingly helpful examples, then rework those into engaging interactions with students. The preservice teachers described a range of problems they encountered, but because the cohort shared no clear images of teaching, our proposed solutions seemed like "one-offs," untethered to any cohesive theory of pedagogy. This meant that values such as equity, inclusiveness, and appropriate challenge were not embodied in principled and commonly understood images of face-to-face work with young learners; rather, they remained aspirations. Our conclusion was that *we had provided the reach but not the footing for our novices.*

We had to reconsider what preservice teachers should learn to do with students, and then help them develop a repertoire of practices to accomplish these valued aims. After months of literature exploration, design, and redesign, we identified a small number of flexible teaching practices and tools that we could focus on together. Currently, twelve practices (four sets of three practices each) make up the pedagogical core of Ambitious Science Teaching, each with its own goals, talk routines, and scaffolds. Over the past decade, an increasing number of teachers across the country, both preservice and experienced, have conducted countless tests of small changes in their classrooms to create effective and context-sensitive variants of these core practices and tools. Today we are excited that this community shares a vision and a common language that allows members to build upon and critique the ongoing work. We regularly incorporate such innovations into our methods courses, learning from K–12 teacher-colleagues across the country as well as from our own experiences.

WHAT WE MEAN BY RESPONSIVE PRACTICE

Our core practices are responsive to learners, meaning they come into being through interactions with students and their ideas. Thinking and doing together in a classroom is always a collective enterprise, cogenerated by bodies, materials, and actions that exist in relation to one another.[4] At the same time, the inherent unpredictability of these environments means that teaching practices are always contingent upon what is happening in the moment, as well as what could happen next; they are in a state of flux, and never a matter of applying some generic technique. As challenging as this seems, our preservice teachers have to first build trustful relationships with students by showing interest in what they do and who they are outside the classroom, listening to and valuing what they have to say, and encouraging cultural as well as epistemic pluralism in the classroom discourse. Following are two main tenets about responsive practices that distinguish them from tasks or moves that can be routinely enacted.

Responsive practices require disciplined improvisation. Well-designed practices require decisions to be made throughout the duration of work being done with learners; consequently no two enactments unfold in the same way. Still, the teacher remains accountable to students' contributions, to the nature of science ideas, and to the norms of the discipline. Sawyer captures the productive tension between principled enactments of instruction and the uncertainty of thinking together with young learners, using the concept of *disciplined improvisation.*[5] He explains that the trajectory of a classroom conversation cannot be reduced to a participant's individual contributions because one cannot know the meaning of their own turn at talk until others have responded. Improvisations, then, are collaborative because no single individual controls what precedes or emerges from them. Facilitating this kind of responsive dialogue can be considered disciplined because it unfolds within broad frameworks, such as Ambitious Science Teaching, that guide practitioner responses to the situation at hand. In relying on such frameworks, a teacher may, at times, address students' needs by extending, adapting, or even discarding elements of a core instructional practice related to the discussion.

Responsive practices are undergirded by principles that prioritize participation, equity, and inclusion. Teachers who use core practices within a professional community depend on shared assumptions about instruction, learning, and educational justice. These both catalyze and constrain the variations that inevitably arise as members of the community seek to modify practices for local needs. For example, these are the grounding tenets of our "collective thinking" practice—better known as whole-class discussion:

- *Broad participation:* For students (including those with special needs, emergent multilinguals, or others historically marginalized in schools and classrooms) to take up academically productive forms of discourse, the teacher must scaffold talk opportunities, reinforce norms of inclusion, and seek out all voices in the conversation.
- *Sensemaking:* Students are encouraged to draw from a wide range of resources to make sense of the activity or topic, including observations they've documented, knowledge from previous instruction, everyday experiences, family stories, spontaneous analogies, and established science facts or theories.
- *Heterogeneity of ideas:* Learners benefit from hearing alternative interpretations of new ideas or activity from peers, culturally diverse ways of reasoning, uses of language, and unexpected sources of uncertainty. There is no one canonical understanding learners must "get to" in the conversation, and the teacher resists privileging Western science norms of knowing at the expense of, for example, indigenous epistemologies.
- *Making thinking visible:* Sensemaking is facilitated by making thinking visible and public via open dialogue and the use of student-created models, charts, lists, marked-up diagrams, realia, and so on.

These tenets do not specify particular moves or activities, but rather they guide enactments that must be sensitive to context as well as content, and to the needs of students in a particular classroom. The decisions will always involve tailoring the practice to accommodate all one's students who are engaging with specific kinds of subject matter, in a specific learning space.

COHERENCE OF PRACTICES AND CHALLENGING THE CULTURE OF "DOING SCHOOL"

Science is not the only subject for which there are core practices. Ambitious Science Teaching, however, as one framework for practice-based teacher education, is unique for two reasons. First, the practices are designed to *work together as a system* to support productive trajectories of learning and participation. For example, the practices used to elicit students' ideas and experiences about a science topic provide formative insights for teachers as they design upcoming lessons. These lessons leverage other practices that expand upon students' ideas through engagement in material activity and disciplinary reasoning. The full repertoire of practices contributes to the larger vision of students working together to use all resources available to them to generate and revise evidence-based explanations over time, and in the process, coming to see themselves as knowers and problem solvers. In addition to the benefits of coherence, research indicates that many of these individual practices have a disproportionately positive effect on children from nondominant communities or those who are otherwise marginalized by schooling.[6]

Second, compared to core practices in other subject-matter areas, Ambitious Science Teaching more explicitly requires preservice teachers to *cut across the grain of traditional pedagogies*, to interrupt the status quo even as they learn the profession. Classroom science in the United States has been characterized as intellectually undemanding, procedural, and disconnected from the development of substantive science ideas.[7] Children at all grade levels are often kept busy copying lecture notes, memorizing vocabulary terms, taking turns responding to known-answer questions, and churning through activities that are only nominally related to authentic science. This is known in the literature as "doing school."[8] Doing school is not just a set of familiar routines; it is a frame that emphasizes teacher control, content coverage, individualism in learning, and students' passive consumption of standardized knowledge. These priorities are so baked into the institutional culture that they are accepted as unremarkable features of classroom life.

This state of affairs is problematic for children but it also impacts how we prepare teachers. For example, clinical experiences for aspiring educators often contradict the methods and approaches advocated in their university courses. And over time, immersion in school settings tends to socialize preservice teachers into normative classroom practices.[9] For these reasons, what we ask of preservice teachers can be complicated by existing expectations in schools and requires a kind of emotional resilience in the face of comments like: "Okay, this sensemaking talk is all well and good, but we don't have time for it, we have to get through this curriculum."

THE CORE PRACTICES OF AMBITIOUS SCIENCE TEACHING

The Ambitious Science Teaching framework (figure 2.1) has been the subject of a recent book that lays out these practices in detail, so we provide only an overview here.[10]

FIGURE 2.1   Ambitious Science Teaching framework

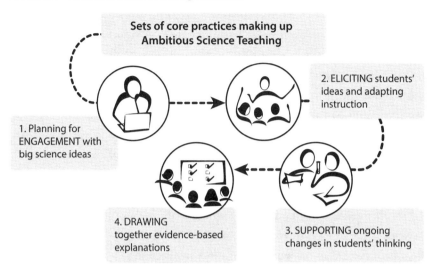

The first set of practices, *planning for engagement with big science ideas*, guides the design of cohesive units of instruction that engage students in constructing and revising explanations for complex phenomena over time. These practices require preservice teachers to select which ideas in the science domain are most necessary for learners to comprehend at deep levels, and to fashion a curricular storyline within which lessons both make sense as individual learning experiences and work together to support cumulative and interconnected understandings.

The first practice in this set of three is identifying the big ideas. If teachers are designing for sixth graders and the topic is sound, they immerse themselves in the standards about waves, acoustics, and the conservation of energy. All relevant ideas are typically transferred to notes and organized on a whiteboard or table. Gradually, the teacher team decides which of these are most fundamental to a deep understanding of sound, and moves them to the center. These concepts will later be woven through multiple lessons in various forms. Science ideas on the periphery are incorporated into instruction as well if they clearly contribute to the idea of sound as way of moving energy from one place to another. Everyone runs into the limits of their content understanding during this process and humility is needed to make progress as a team.

The second practice is selecting an anchoring event and an essential question. Good anchoring events require students to integrate a number of important science ideas together in the process of explaining them. Anchoring events can have historical significance, such as the selective survival of particular species of finches on the Galapagos Islands. They can be about issues of social justice, such as the causes and consequences of sea-level rise for those living in poverty-impacted coastal communities around the world. They can be about epic phenomena like supernovae or everyday occurrences like applying chemical products to your hair. These events should be culturally accessible by all students and drawn from experiences that students from nondominant communities find relevant and engaging. Essential questions further define the challenge for students. Occasionally learners themselves come up with better ones than teachers do; for example, in a seventh-grade unit on solar eclipses, one boy asked why they were so predictable. This, the teacher recognized, was more intellectually demanding that just explaining why eclipses happened, and agreed with her students that this should become the focus of the next few days of investigation.

Finally, in the third practice, teachers start by writing a gapless explanation for the anchoring event and break it down into its constituent ideas. "Gapless" means that the teacher, and later the students, try to include all the links in the causal story. Once the narrative is written, one or more lessons are selected to explore each of the contributing ideas and they are sequenced in a way that supports students in building explanations and scientific models over the course of the unit.

The second set of practices is referred to as *eliciting students' ideas and adapting instruction.* Here teachers begin units by finding out what resources students bring to the topic, including the informal language they use to talk about the focal phenomenon, their everyday experiences, and partial ideas and puzzlements. For a ninth-grade unit on how greenhouse gases affect climate change, a teacher might show an image of bumper-to-bumper traffic on a freeway next to a graph of historical levels of atmospheric carbon dioxide, and begin by asking, "What do you see?" Because sensemaking talk by students is central to our core practices, we scaffold the dialogue by sharing what counts as an observation, giving quiet writing time after the question, sharing ideas with a partner, and then opening the discussion by explicitly inviting stories about outside-of-school experiences and drawing on children's funds of knowledge related to vehicles, combustion, or pollution in general.

In a few minutes, teachers transition from observations to hypotheses, asking, "What might make the line on the graph look like this?" No technical vocabulary is used as the teacher and students cogenerate a list of speculative hypotheses and partial ideas. Often students are asked to draw models of the phenomenon, including what is unobservable. Making thinking visible in these ways— through conversations, lists of rough-draft ideas or hunches, initial models —allows students to hear alternative ways of reasoning by their peers, to prioritize different explanatory ideas, and to assess their usefulness. At the end of the day, the teacher quickly surveys the lists, models, or exit slips that students have generated and uses these to modify upcoming lessons based on what their current ideas and questions are.

The third set of practices, *supporting ongoing changes in students' thinking*, involves cycles of learning activities followed by the public representation of new ideas and collective reasoning by class members about what they've learned and how it informs their thinking about the anchoring event. Students repeat these experiences multiple times during the unit.

The initial practice in this set is a form of interactive direct instruction. This surprises a lot of educators who initially cannot reconcile sensemaking with being told about ideas. However, many powerful science concepts are abstractions that cannot be "discovered" by students through any form of material or data-based work (think of alleles or chemical equilibrium). These ideas have to be presented by teachers, not as memes to be memorized but as tools to reason with. For third graders who are studying how one apple tree can produce others in an open field, a teacher might introduce the idea of pollinators (bees, birds, moths, ants, etc.) and what these creatures seek out as they visit flowering plants. After a read-aloud about bees and conversations where students pose their own questions about the role of these insects, the teacher can place them in groups of three and give them a map that shows the daily comings and goings of bees around a stand of apple trees. Children are asked to record patterns they notice. The teacher then visits with each group for a few minutes, listens to what they are saying, prompts them to clarify ideas, checks whether they understand one another's contributions, and invites everyone to speak. Before the teacher leaves, she may encourage groups to share provocative or unique ideas later in whole-class discussion. Emerging multilingual students benefit from this extra time to compose what they will report out.

Later, students regroup as a whole class to make sense of the activity together. They negotiate responses to three questions that are similar for any grade level and any topic: What patterns or trends did we observe?, What do we think caused these patterns?, and How does this activity help us understand the anchoring event? To publicly display the group's ideas, these conversations have a structure and a tool (called a summary table) that allow more students to participate than in typical discussions, and in a wider variety of ways. Students get access to the reasoning of their peers and can acquire new forms of argument as they move from legitimate peripheral participation in the dialogue to playing a central role as the weeks go by.

This set of teaching practices helps students learn to marshal various resources for solving problems and developing new knowledge (e.g., science practices, data, graphs and other representations of information, one another's ideas, authoritative science ideas). They also support the development of students' academic discourse.

The final set of practices, *drawing together evidence-based explanations*, supports students in using ideas and information from preceding activities to revise their current explanations and scientific models. This set is typically used in the middle of the unit and once again near the end of the unit. In using this set of practices at the end of a unit, students and the teacher begin by coconstructing a gotta-have checklist. They deliberate together and come to agreement about this question: "What has to be included in a credible and complete explanation?" In a seventh-grade classroom where students are trying to understand why solar eclipses are rare yet predictable, they can offer a variety of ideas. Some mention the relative sizes of the earth, moon, and sun, while others suggest that the "off-kilter orbit of the moon" has to be somewhere in the explanation. One student puts in a bid for the time it takes as the earth orbits the sun, but most of his peers push back, saying that no activities, readings, or conversations indicated that this mattered. As the teacher records and lightly edits this list, she may make an executive decision and ask: "What about the role of gravity that we explored in the computer simulation?" Several students affirm that this should be included on the checklist; however, other students remind the teacher that they had talked about the relationship between mass and gravity a few days earlier, but never came to a conclusion about whether it was relevant to the eclipse story. She registers their concern and offers to do a quick revisiting of this idea.

Later, as part of a culminating performance assessment, the teacher asks students (individuals or pairs) to make final versions of their drawn models for the anchoring event as well as produce written explanations. Students are encouraged to examine records of their thinking from earlier in the unit that are posted on the classroom walls. These include revised lists of hypotheses, summaries of how they made sense of different activities, and samples of mini-models drawn during labs.

As with other teaching practices, students with special needs are supported through instructional designs that can flex to allow for extra time, verbal directions alongside written ones, and cues from the teacher to help them focus on key parts of the task. Emerging multilinguals can be partnered with peers who speak the same home language, but are slightly more proficient at English. These and other supports enable everyone to show the most of what they know by working across languages and using a broader linguistic and extralinguistic repertoire to communicate. The teacher circulates and presses students to show where they've included various ideas from the class's gotta-have checklist, and asks them to describe how it contributes to their explanation. This set of practices, as an ensemble, reveals the depth and generalizability of what student have learned. At the same time, it provides feedback to the teacher about the efficacy of instruction, including tools and routines that have to be modified to serve students better.

Practice-based approaches to teacher education, and in particular the Ambitious Science Teaching framework, accomplish goals for student sensemaking and inclusion by bringing together elements of an emotionally supportive environment, productive discourse, engagement with the discipline, collaborative work, scaffolding, the appropriate level of intellectual challenge, and attention to equity—elements that some teacher preparation programs address in isolation from one another or discuss in abstract and disembodied terms. Increasing the prominence of student voice is central to this framework. We want young learners to take responsibility for decisions about phenomena they'll study and to decide how topics can be made more relevant or could address issues they are concerned about. Students should also have a voice about instruction itself, providing regular feedback to teachers on activities or norms that seem to be productive for them and others that seem to hinder sensemaking or their sense of belonging in the classroom.

Learning to do this work takes time. This is why many of our methods classes are devoted to modeling, rehearsing and critiquing these practices, and then unpacking the social contexts within which this work is done. What do we leave out? Learning how to use microscopes, reading biographies of scientists, working with critters in the lab, formatting lesson plans, and running through generic curriculum activities. We do give time to learning about standards, but we don't obsess about aligning everything we do with them. Standards, and the reform rhetoric that goes with them, have never changed who gets to participate in classrooms or how. We argue that courses like methods should be grounded in understanding the conditions under which all students can learn, and the supports and opportunities teachers must provide every day for expansive forms of sensemaking. This requires a letting go of safe activities and bureaucratic exercises in methods classes that have only sketchy connections to children's development or well-being. Unless we prioritize the student experience, we risk apprenticing our novices for "doing school."

DO AST CORE PRACTICES IMPACT THE FIRST YEARS OF TEACHING?

Over recent years we have followed multiple cohorts of preservice teachers beyond their work in preparation programs and into their first year of teaching. What we found is encouraging and at the same time compels us to acknowledge that we need more data. In one study, Kang and Windschitl wanted to know how two groups of beginning teachers—one that graduated from a core practices program and a matched group who were not taught core practices—compared in their abilities to provide their students with varied opportunities to learn.[11] A total of 116 science lessons taught by forty-one first-year teachers were observed and analyzed. Among the 77 lessons taught by teachers from the core practices program, about 20 percent of these were categorized as (1) rich in opportunities for students to engage in scientific practices, (2) aimed at authentic and complex problems, and (3) high in teacher responsiveness to students' ideas. In contrast, only about 2.5 percent (one of 39) of lessons in the comparison group were categorized as high in these areas that supported students' opportunities to learn. In about 90 percent of the lessons taught by the comparison group, students mostly experienced science as learning of facts, procedures, or correct explanations. The authors concluded that core practices supported preservice teachers in formulating an actionable curricular vision, helping them notice, attend to, and build upon students' ideas in classrooms. They cautioned, however, that emphasis should be placed on the vision and pedagogical goals underlying the core practices, rather than the ungrounded use of strategies or tools.

In another study, Thompson, Windschitl, and Braaten examined how twenty-six first-year secondary science teachers developed their beginning instructional repertoires while working in either a school community infused with visions and tools supportive of ambitious teaching or one that reinforced traditional practices.[12] Two-thirds of the preservice teachers in this study were able to develop forms of responsive and equitable practice despite working in school environments with standard or conservative teaching practices.

In more recent case studies of three teachers prepared with core practices and tools, Kang and Zinger found that providing high-quality learning opportunities depends on many factors: the time teachers have to plan and to reflect on instruction, encouragement for responsive pedagogy provided by colleagues, and for some preservice teachers, the negative impacts of cultural scripts about teaching they acquired as science students themselves.[13] These interacted to facilitate or undermine their responsiveness to students and whether they provided opportunities to learn on a daily basis. The authors found that using core practices as the main curriculum of science methods courses could help prepare preservice teachers for equitable teaching if the approximation of these practices allowed them to question their normalized views, expectations, and practices around disciplinary teaching and learning. Core practices by themselves are limited, however, in their ability to develop preservice teachers' critical consciousness about racism and systemic inequities. We take up these ideas further in the next chapter as we consider equity principles that are foundational to the profession.

The practice-based teacher education community has continuing work to do. We are constantly reevaluating the practices and our own pedagogies to help novices enact instruction that is fundamentally different from traditional schooling. Driving some of this change is the self-critique by members of our own community, including the authors of this book. Every contributor brings specialized knowledge to bear on the challenges of implementing core practices and the implications for preparing educators to work in systems that often both nurture and oppress children. Some of our colleagues, for example, do research on how teachers use social networks to spread innovative strategies; others study the construction of assessments that can motivate marginalized students and allow them to show the most of what they know. Still others run science clubs for urban girls or investigate how teachers make use of indigenous students' everyday knowledge and epistemologies in their curriculum. All these projects help situate core practices within discussions about broader and more important goals for children and their families. The influx of ideas and new questions from these varied areas of inquiry can keep us from becoming too inward-looking about the practices or feeling settled about what we have accomplished so far.

Every aspect of Ambitious Science Teaching and the use of core practices is open to inquiry. This includes questioning the efficacy of teacher educator pedagogies, exploring how responsive practices get implemented or set aside during the clinical experience, and challenging the foundational assumptions we make about whether our practices actually embody a more expansive and socially just view of professional learning. While we embrace any and all critiques, we are also about action. This is a rare historical moment of relative unity and coherence among teacher educators, as we are often separated by our academic specialties and occasionally by the ways we use language. A robust vision of compassionate practice can enable scalable reforms in how we prepare science educators, and shape broader conversations about classroom conditions that allow more children to make deeper sense of the natural world and their place within it.

CHAPTER 3

Culturally and Linguistically Sustaining Approaches to Ambitious Science Teaching Pedagogies

JESSICA THOMPSON, KIRSTEN MAWYER, HEATHER JOHNSON,
DÉANA SCIPIO, AND APRIL LUEHMANN

In this chapter we use a critical equity perspective to examine the Ambitious Science Teaching (AST) core practices and pedagogies with the aim of supporting teacher educators in grounding learning opportunities in students' cultures and identities and developing culturally and linguistically sustaining practices.[1] Gutiérrez suggests that, traditionally, scholars and practitioners have described equity in terms of issues of access, opportunity, and achievement, and that a critical perspective must also attend to issues of power, identity, and the sociopolitical context.[2] The Ambitious Science Teaching vision is that science classrooms and learning spaces can be places where students experience science in ways that have relevance and power in their own worlds and cultures, especially for students from groups who have been historically marginalized. This commitment translates to a need for teachers to be continual learners of students' local culture and identities, and to be brokers of science as they support learners in understanding how humans have constructed science and the power and status issues related to who determines what, how, and why science is accomplished.

With knowledge of the discipline and how it has been constructed, children and youth can begin to reshape a new science that is not only culturally sensitive and responsive, but that also helps sustain nondominant community participation.

Our hope is that this chapter represents a starting place for meaningful dialogue about how to embed a critical perspective in preservice science teaching methods courses and in teacher education programs more broadly. In terms of positionality, we (the authors of this chapter) are all science teacher educators who have used Ambitious Science Teaching practices in partnerships with K–12 classrooms, schools, and districts and in out-of-school science learning environments, and in our teacher education contexts. Our teacher preparation programs vary; some of our programs have mostly white preservice teachers while others have over half as preservice teachers of color. Some of our science teaching methods courses are situated in programs that deeply consider issues of social justice, while others are in the process of trying to reorganize along these lines. In this chapter, we describe critical equity principles that guide our pedagogies in our science teaching methods courses in order to support teachers in desettling expectations, navigating interculturality and intersectionality, and teaching science in ways that view racialized identities as assets for science learning and identification with scientific domains.[3] We examine four critical equity principles: (1) Recognizing our own and others' worlds and developing critical consciousness, (2) learning about and prioritizing students' communities and cultures, (3) designing for each student's full participation in the culture of science, and (4) challenging the culture of science through social and restorative justice. We begin with a vignette that applies the four principles in complex practice to show how they work together to support preservice teacher learning; we then unpack the principles in the following sections. This case focuses on a teacher education program that aims to sustain and revitalize language and culture in Hawai'i, recognizing that culture and language are ways of interpreting the world that are constantly in flux, and when placed at the center of teacher education work, can offer transformational learning and change.

In Hawai'i, Kirsten is developing a science teacher education program in partnership with the Hawai'i Department of Education and is undertaking the important work of preparing preservice teachers to act as agents of change; the program aims to revitalize a community as much of the culture and language has been erased by European colonization. In 2015, the superintendent stated a vision for moving forward: "When I walk into a Hawai'i public school, I want to close my eyes and know that I am in a school in Hawai'i . . . and not somewhere else."[4] When schools and the teaching and learning that goes on in them are divorced from place, it "limit[s] experience and perception; in other words, by regulating our geographical experience, schools potentially stunt human development as they help construct our lack of awareness of, our lack of connection to, and our lack of appreciation for places."[5]

This program uses Ho'okupa Hawai'i, a series of teacher pedagogies designed to help preservice teachers imagine and develop a vision of teaching practices that will support the design of meaningful, relevant, and culturally sustaining and revitalizing learning experiences. Preservice teachers are taught to consider context—the students, the school, the community, and the place, as well as the layered histories and narratives of that place over time—as they craft learning experiences and curriculum. The starting point is to engage preservice teachers in the principle of *recognizing our own and others' worlds and developing critical consciousness.* Each preservice teacher is challenged to unpack their identity by exploring the question, Who am I? Reflecting on this question encourages and supports preservice teachers to bring their own cultural positionalities into explicit visibility and to identify how they relate to issues of equity, diversity, access, and power. Exploring this question provides candidates important insights about the forces that shape their cultural beliefs, values, norms, and identities, such as, "The homogenous dominant culture I was living in spoke a message to me: I was intrinsically *wrong*" and "Growing up in Hawai'i has also opened my eyes to a unique cultural diversity that has given me the opportunity to learn about multiple cultures." Also, the question provides a space for thinking about how to agentively and sensitively negotiate the cultural, historical, political, and physical landscape(s) they will inhabit as teachers, as this comment from a preservice science teacher in Hawai'i's program shows:

> My family was never prevented from practicing our culture. I enjoy the *hula* [dance], the *'ōlelo* [language], the *kalo* [food staple] fields, the *loko i'a* [fishpond], conservation, education, volunteering and perpetuation. The blend of ancient practices and modern practices causes confusion in the students. They need resource people and knowledgeable mentoring people, in order for them to make sense of their world yet also retain their opportunities to *mālama* [to care for, preserve] their culture.

The "Who am I?" question was also a jumping-off point for considering the sociohistorical context of settler colonialism, which displaced Hawaiʻi's indigenous *Kanaka ʻŌiwi* (Native Hawaiian persons) from their lands, language, and culture. Moreover, colonialism's legacies continue to subtly and not-subtly privilege "whiteness" and practices of the dominant culture across contexts that extend into the present day and shape the context in which preservice teachers are teaching.[6] One preservice teacher noted: "I am aware of the long and complicated history of modern Hawaiʻi; the white missionaries condemned and denied land, language, and fair education to Hawaiian people and then to immigrants. As a result, there are many people now in Hawaiʻi that distrust white people's motivations and way of doing things."

Another key component of this principle is for preservice teachers to deepen their understanding of the students and the context in which they teach because this information directly influences the decisions they will make about planning, instructing, and assessing students. As part of the work candidates explore these questions: Who am I teaching? Who are my students? The goal is to promote their development as culturally sustaining and revitalizing educators who create intellectually safe and inclusive learning environments that support individual and collaborative learning, encourage positive social interaction, and promote active engagement in learning.

The next part of the project asks candidates to explore the equity principle of *learning about and prioritizing students' communities and cultures*. Before they can design meaningful place-based curriculum and learning opportunities, preservice teachers need to gain a deeper understanding of the community in which their school is situated. Preservice teachers created community asset maps and inventories of their classroom, school, and community. To explicitly attend to the sociohistorical context and culture of Kanaka ʻŌiwi, preservice teachers are also asked to map and inventory the *ahupuaʻa* (land division that extends from the mountains down to the ocean) the school belongs to, how the water moves through it, and key *wahi pana* (geographic, historic, and cultural places of interest). This is essential because in Kanaka ʻŌiwi culture, *hahai no ka ua i ka uluālāʻau*—the rain follows after the forest, which points to the critical ecological, practical, and spiritual roles *wai* (freshwater) plays as it flows through *ahupuaʻa*, sustaining and feeding the entire community.[7]

After mapping and inventorying their community, preservice teachers interviewed cultural practitioners, *kupuna* (elders), representatives of place-based organizations, scientists, and other community stakeholders to identify an injustice the community wanted addressed. They used the *mana'o* (thoughts) and *'ike* (knowledge) of community members to help them think about how to leverage community assets to design a place-based science learning experience that engages students in thinking about social justice and dimensions of place, and that strengthens their sense of belonging.

The final step of the project specifically asks candidates to explore that question: How can you use community assets to develop a place-based science learning experience that engages students in thinking about social justice and dimensions of place, and strengthens their sense of belonging and relationship to Hawai'i as they learn a big science idea? Designing place-based Ambitious Science Teaching that empowers students to use science to connect to and understand real-world phenomena in the place they live requires candidates to operationalize the principles of *designing for each student's full participation in the culture of science* and *challenging the culture of science through social and restorative justice.* One preservice teacher described this work as "envisioning a place where students can get their hands dirty protecting and restoring cultural and ecological resources, see the results of their hard work, and feel pride in their contributions." She took her students on a *huaka'i* (learning journey) to a national park where they were able to see and hear the bird songs of endemic *'apapane* and *'i'iwi*, which are endangered honeycreepers, and engage with field rangers and scientists to work on propagating an endangered plant found only within the park. She reflected that "by participating in the efforts to save rare, endangered species, students develop a vision of their purpose and path toward the future."

There is a *'ōlelo no'eau* (proverb), *A'ohe pau ka 'ike i ka hālau ho'okāhi* (all knowledge is not taught in the same school), that embodies the importance of and need for multiple learning spaces, perspectives, and sources of knowledge that extend beyond the traditional classroom.[8] Developing place-based science curriculum opened up a space for critiquing, challenging, and changing strictly European notions of science and school science to include other ways of knowing that in turn help *all* students understand and make sense of the world.

In the following sections we unpack each of the equity principles, describe teacher educator pedagogies we use in our coursework to decenter whiteness, attend to and disrupt traditional power structures, and create learning environments that recognize multiple ways of building identities as learners, scientists, and publicly engaged citizens.

PRINCIPLE 1: RECOGNIZING OUR OWN AND OTHERS' WORLDS IN DEVELOPING CRITICAL CONSCIOUSNESS

As described in the Hawai'i case, preservice teachers and the teacher educator began with an often uncomfortable consideration of self. They examined their own positionality and biases and then began to consider how to take action against oppressive aspects of schooling and society once they were aware of such injustices. Developing critical consciousness requires looking inward and recognizing our own identities and cultures while respecting and learning about students' identities and backgrounds such that we can develop grounded, meaningful, and transformative learning opportunities with students. In our teacher education courses, we have found that this begins with a careful examination of our own cultural frameworks and identities. As most of the teaching force is white, female, and middle class, this means widening our perspectives to include a study of whiteness, the culture of power, the sociohistorical context of settler colonialism and how it shapes school and science learning, and the ways in which our positionality shapes our narratives about race, class, language, and culture in science classrooms and informal learning spaces.[9] Furthermore, because preservice teachers and teacher educators have grown up learning science in unjust and racialized contexts and often have not been positioned to question these perspectives or values, we often accept the epistemology of Eurocentric ways of knowing. Thus, our science teaching methods courses need to provide preservice teachers experiences that challenge Eurocentrism as the mainstream so we can start to shape new visions of teaching and learning science that will be more inclusive of the rich linguistic, racial, and cultural resources that students will bring into the learning environment.

We have found that consciousness work with preservice teachers requires a historical deep dive into one's own K–12 experiences with science in and outside of school and into the dominant and unbalanced narratives that are often portrayed in science learning environments. Like many teacher educators, at the beginning of our science methods courses we ask preservice teachers to reflect on their past experiences in K–12 science. Here we share one example of writing cultural K–12 autobiographies. We ask preservice teachers to reflect on the way in which they experienced continuity (or not) among the worlds of home, school, and peers and write a self-narrative about cultural and linguistic connections and disconnection, identity, access, and power in school science.[10]

Along with the autobiography, preservice teachers can explore their identities through alternative formats such as cultural mosaics using cutouts from magazines or digital images to represent their personal identities and connections to science (see figure 3.1).[11] Preservice teachers will have diverse experiences include some who have experienced oppression as well as others whose worlds were fairly aligned. Teacher educators will want to recognize these differences as strengths and resources, and acknowledge that this is deep identity work that may cause people to shut down or push back. Frequently revisiting these narratives, in light of experiences in classrooms and additional readings, can help preservice teachers broaden their narratives to include critical consciousness about issues of privilege, power, problematic dichotomies, and perspectives on emergent bilingual students.[12]

More than just doing assignments such as these deep dives into Ambitious Science Teaching experiences, we have found it important for preservice teachers and teacher educators to continually ask questions like these: How do I participate in the culture of power? Whose knowledge is valued in the classroom/learning space? What counts as scientific knowledge? What kind of science identities do I want to promote and reinforce? These questions hold true for the science methods course and K–12 classrooms.

PRINCIPLE 2: LEARNING ABOUT AND PRIORITIZING STUDENTS' COMMUNITIES AND CULTURES

In the Hawai'i case, preservice teachers were charged with learning the places, people, histories, and values of the community that was home for the students they would eventually teach. Especially when preparing preservice teachers to teach students from communities and cultures that differ dramatically from their own, we have developed an appreciation for the importance of protecting time and space for preservice teachers to identify, recognize, and learn with and from the expertise in the community. In so doing, not only are they able

FIGURE 3.1

to develop a sense of local resources that can be used to teach and learn science; more importantly, they are given opportunities to develop an appreciation for the beauty, strengths, and history that science education might celebrate and sustain as well as a sense of potential community needs that science education might serve to address. As outsiders, preservice teachers need opportunities to access community knowledge, develop an appreciation for the local context, and understand the resources available in the community, while also pushing back on deficit views of "urban" or "rural" communities. We have explored various ways to structure these learning opportunities for preservice teachers, including asking them to design and conduct original science investigations situated in local places (e.g., farms, streams, playgrounds), conducting brainstorming sessions with an advisory panel consisting of diverse community stakeholders around potential local anchoring phenomena, and constructing an asset map of a community (see figure 3.2).

Creating a community asset map of the areas (neighborhoods) surrounding a school allows preservice teachers to gain a deeper understanding of the communities with which the school intersects. The goal is to ensure that all preservice teachers understand and adopt the fundamental assumption that good things exist in every neighborhood, and science classrooms can highlight, build on, and contribute to these assets.

To create a community inventory and asset map, preservice teachers walk the neighborhood to identify what they observe. To add perspective, they take the same walk with a member of the community who shares stories of lived experiences and their own observations about the spaces, places, and people in the community. Enlisting the support of a "local tour guide" helps preservice teachers see what is not immediately visible, and provides an experience of adopting the lens of an eager and invested learner. As they walk together, the preservice teachers look for and document a variety of community resources, including physical assets (e.g., buildings, transportation), economic assets (e.g., businesses, restaurants), stories that offer insights (e.g., memories, values), local residents (e.g., youth, artists, pastors), and institutions and associations (e.g., churches, hospitals, youth clubs). Working with students and local stakeholders to create a community inventory and asset map can realize benefits that extend far beyond accumulating knowledge of and about particular locations; the sharing of stories in and through the walk and mapmaking can nurture relationship building

FIGURE 3.2 East High community asset map

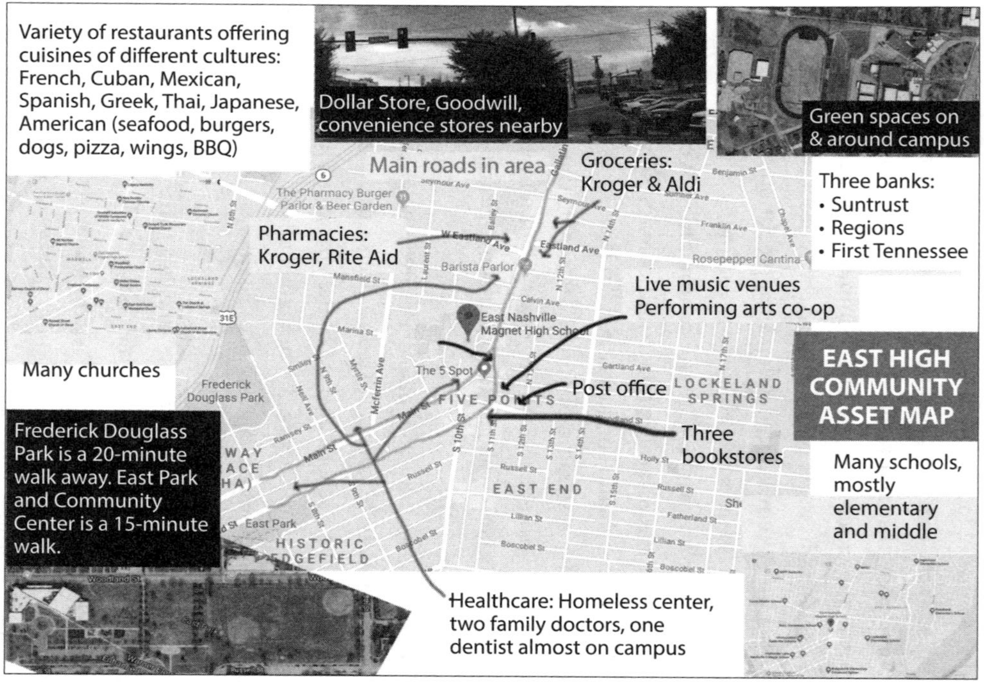
Variety of restaurants offering cuisines of different cultures: French, Cuban, Mexican, Spanish, Greek, Thai, Japanese, American (seafood, burgers, dogs, pizza, wings, BBQ)
Dollar Store, Goodwill, convenience stores nearby
Green spaces on & around campus
Main roads in area
Groceries: Kroger & Aldi
Three banks:
• Suntrust
• Regions
• First Tennessee
Pharmacies: Kroger, Rite Aid
Live music venues
Performing arts co-op
East Nashville Magnet High School
Many churches
Post office
EAST HIGH COMMUNITY ASSET MAP
Frederick Douglass Park is a 20-minute walk away. East Park and Community Center is a 15-minute walk.
Three bookstores
Many schools, mostly elementary and middle
Healthcare: Homeless center, two family doctors, one dentist almost on campus
FIVE POINTS
LOCKELAND SPRINGS
EAST END
HISTORIC EDGEFIELD
Frederick Douglass Park
East Park
The Pharmacy Burger Parlor & Beer Garden
Barista Parlor
The 5 Spot
Rosepepper Cantina

that serves to sustain scientific inquiry and inform local science education in powerful and ongoing ways. Once preservice teachers identify existing assets of the school and the community, they summarize how they can leverage the assets as resources to inform the selection of an anchoring phenomenon, elicit student connections to the phenomenon, and support students' changes in scientific thinking about the phenomenon, as well as how they can give back to honor and enrich the local communities. While preservice teachers might do this walk once as part of a social foundations course, it is important to revisit and add to the map in light of upcoming science units of study.

One of the first decisions teachers make when they use Ambitious Science Teaching practices (outlined in chapter 2) is to plan to engage student thinking by selecting a worthwhile phenomenon, with the explicit goal of choosing one that has meaning to students, to both anchor scientific learning and connect to grade-level standards (AST core practice set 1). Pedagogical decisions in the other three sets of AST practices are also strengthened by being grounded in and contributing to local culture, (AST core practice set 2), counting personal stories as additional evidence for scientific reasoning (AST core practice set 3), and inviting community stakeholders as authentic audiences to motivate evidence-based explanations (AST core practice set 4). Students, parents, and local community members hold needed expertise for science education on what phenomena have relevance, what community experiences and priorities are shared, and what local needs science might address. Preservice teachers need such opportunities to come to know their students and their families both personally and academically. Hanging up posters of people of color in science or referencing other ethnic cultures in story problems is inadequate for creating a learning environment that positively and effectively contributes to the diverse cultures that make our world strong and vibrant. Learning with and about students, families, community, and local culture occurs through relationships.

While doing something like a community asset map is helpful for identifying available community resources, the real work of learning about and prioritizing students' communities and cultures is accomplished through continually asking important questions, such as: What types of teacher-student, student-student, teacher-parent, and teacher-community relationships are supportive of student learning? How can teachers build bridges between multiple cultures represented in the school community? How do teachers attend to and leverage language, race, and culture when selecting phenomena and designing engaging and inclusive curriculum with historically marginalized students?

PRINCIPLE 3: DESIGNING FOR EACH STUDENT'S FULL PARTICIPATION IN THE CULTURE OF SCIENCE

To prepare teachers to teach in multicultural and multilingual classrooms, we must support preservice teachers in attending to equity at the level of classroom interactions, practices, and design and focus on students' languages and cultures as resources for learning and sensemaking in science classrooms. We have found that preservice teachers can develop classroom cultures that value students' resources by reconfiguring their understanding of interpretive power.[13] Interpretive power is a reflective teacher practice that builds upon noticing in classroom environments. It refers to "teachers' attunement to (a) students' wide-ranging sense-making repertoires as generative intellectual resources in science learning and teaching, and (b) expansive pedagogical practices that encourage, make visible, and intentionally build on students' ideas, experiences, questions, and perspectives on scientific phenomena."[14] Developing interpretive power connects to ideas discussed in equity principle 1 and asks preservice teachers and teacher educators to do the personal work necessary to recognize that *we* are the ones who need to become flexible and learn to recognize how students' ideas and contributions are generative and scientific. Developing interpretive power challenges us to recognize that students are always making sense, and we must attune ourselves to their sensemaking practices.

While we see the value of connecting curriculum to students' lives to make learning authentically meaningful and purposeful, we also recognize the potential generative nature of this type of learning for deepening science understanding. Determining what counts as scientific for teachers and students in multiple contexts is a central part of creating curricula at the intersection of science and society. Preservice teachers need to both explicitly teach how Eurocentric cultural perspectives are intertwined with science and shift curriculum and classroom communication to include students' views and communication patterns. For example, Kirsten has her preservice teachers in Hawai'i experience an Ambitious Science Teaching unit as students; students learn about moon phases by concurrently looking at *kaulana mahina* (the Hawaiian lunar cycle), which groups the thirty moons into three periods of ten days next to the eight Western phases. Running these science narratives in parallel helps privilege multiple ways of knowing scientific phenomena.

As teacher educators, we are invested in developing ways for preservice teachers to have experiences as learners that position them to understand science through the lens of culture and language. One of the ways we emphasize the importance of drawing on students' languages is to model this in our teacher education courses. At the outset of the course, we research which languages preservice teachers speak and then incorporate these languages into lessons where we model the teaching of science. In some cases, we simply translate key concepts into the multiple languages, then open up a conversation with preservice teachers about the ideas and how they do or do not translate, and what kind of power the concepts have for learning the science ideas. In other cases, we do a bit more research on the function of the language. For example, when modeling a fourth-grade unit on sound, we introduced the words *ola* and *oda* as two terms for "wave" in Spanish. The language is powerful since fundamentally the terms help distinguish longitudinal and transverse waves. Similarly, when modeling a unit on forces and motion and discussing how to create bridges between everyday and scientific language, we ask the students to draw a Venn diagram with the term *velocity* in the middle. We have students think about this in English and in Spanish, since the terms in English can have very different connotations, but in Spanish velocity translates to *velocidad*, so the everyday use is not as distant from the scientific meaning. By privileging multiple languages in methods courses, we help preservice teachers understand the importance of linguistic diversity and see how different languages might provide access to scientific ideas.

A focus on sensemaking within multilingual and multiliterate science learning communities helps preservice teachers consider language as a resource in science classrooms. This allows preservice teachers to think about whose voices are incorporated into classrooms, and how classroom communities can encourage multiple ways of knowing as a way to design learning environments and expand preservice teachers' ideas of what full participation in a science classroom looks like from a student-centered perspective. Learning to see students' languages as repertoires of practice in science classrooms provides opportunities to think about how students use language functionally in order to accomplish work. For example, naming translanguaging as a legitimate form of participation in sensemaking in science classrooms challenges teachers to trust that students who are partnered with others who speak the same home language are engaged.[15] Questions of language use in the science classroom are further complicated by entrenched ideas of science as a body of knowledge that requires the use of specific vocabulary. Thus, we have found it important to help preservice teachers question (1) if and how vocabulary can become a gatekeeper to science participation if teachers focus on English language proficiency above sensemaking practices and (2) how the practice of front-loading vocabulary sends a message that there is a "right way" to participate in science classrooms versus how co-constructing ideas opens up opportunities for sensemaking. Such explorations help preservice teachers develop interpretive power and see the science in their students' contributions.

To help preservice teachers analyze student sensemaking and develop their interpretive power, we as teacher educators engage preservice teachers in analysis of video and transcripts of interactions in their classrooms. Classroom interactions are fleeting moments coconstructed by students, resources, discourse, gesture, and curricula. In the moment, it can be quite difficult for teachers or teacher educators to deeply understand the sensemaking practices students are using. In one teacher education program, we have come to focus on equity in practice by asking preservice teachers to select and analyze moments when their students were engaging in sensemaking in their science classrooms. Preservice teachers capture short videos (three to five minutes long) of their classrooms and create transcripts of the discourse. Moments that focused the camera on students rather than teachers are the richest for the purposes of this exercise. Preservice teachers are asked to contextualize the videos by bringing in examples of student work and lesson materials. Working in small groups, preservice teachers introduce a clip from their classroom and watch each video at least twice. The first time they watch the clip all the way through, taking a moment at the end to ask clarifying questions. Then they watch the video a second time, but pause it to discuss moments of teaching and learning. Any viewer can call "stop" to point out something they noticed or to ask a question.

Finally, we use the transcripts of the video to focus our attention on the ways that students are constructing knowledge and using multiple sensemaking and language practices. For example, if a student is using their home language to figure out a construct, then the language context not only helps the teacher make sense of what's going on, but also gives the teacher a future tool for use with that student. During discussions teacher educators keep the conversations centered on understanding how students are developing and communicating explanations of scientific phenomena, demonstrating their understandings, or interacting with the ideas of other preservice teachers. The combination of video and transcript work focuses attention on sensemaking likely to be missed in classroom interactions and allows teachers to develop their interpretive power.

To keep this principle alive, preservice teachers and teacher educators can continually ask questions like these: How are culturally and linguistically sustaining practices more than just good teaching? How can I apply Ambitious Science Teaching practices and pedagogies in multiethnic/multilingual classrooms? Is it my responsibility to learn how to speak my students' home languages (even just a little)?

PRINCIPLE 4: CHALLENGING THE CULTURE OF SCIENCE THROUGH SOCIAL AND RESTORATIVE JUSTICE

As preservice teachers are introduced to Ambitious Science Teaching core practices, there is an opportunity to reimagine classrooms and the enterprise of science through a social and restorative justice lens that conceptualizes science teaching and learning as sitting at the intersection of identity, diversity, justice, and action.[16] These four domains partnered with AST core practices provide students the knowledge and skills needed to reduce prejudice, advocate for collective action, and repair harm as they participate in the discursive practices of science both within and outside of the classroom. Embracing restorative justice provides equitable learning environments, avoids punishments, nurtures healthy relationships, repairs harm, and transforms conflict in a manner that supports and respects the inherent dignity and worth of all.

To challenge the culture of science and school science, we as teacher educators have designed opportunities for preservice teachers to learn how to simultaneously help students participate in the culture of science while critiquing the current and historical ways school science has failed to be inclusive and has replicated patterns of injustice. Students need opportunities to question and challenge myths about "science as truth" and respectfully engage a heterogeneity of ideas and multiple ways of knowing. They need to be able to see and critique European science as a dominant way of knowing and see how the incorporation of other ways of knowing might change the discipline. Social justice is the goal, meaning that all students become scientifically literate, become empowered to make decisions in their own lives and communities, and learn to address current and historical inequities in society through science. This type of social justice work inherently involves identity work, on the part of both teachers and students. In particular, students need spaces to author who they are, their ways of knowing, and their perceptions of justices and injustices in the world in which they live. Furthermore, they need support in leveraging these ways of knowing alongside traditional Western ways of knowing and doing science.

One way for teacher educators to establish an intellectually safe space with their preservice teachers is to start by having everyone share personal examples from their own educational experience when they either did or did not feel intellectually safe and draw on those experiences to reach a shared understanding of how members will work with one another throughout the methods course.[17] What develops out of this sense of community is a growing trust among participants, the courage to present one's own thoughts, and the opportunity to practice democracy in a fundamental way. Preservice teachers can use this approach with their own students to establish intellectually safe spaces in which different ways of knowing and understanding science phenomena are valued as resources.

Attending to social justice and restorative justice calls for preservice teachers to gain practice in identifying injustices. As products of school systems that have perpetuated injustices, preservice teachers may find them difficult to see. One activity is to have students do a "deep dive" into a particular injustice. This begins with having preservice teachers work with students, teachers, and community members to identify an example of unfairness on the individual level (e.g., biased speech) and/or injustice at the institutional or systemic level (e.g., discrimination) that the school and/or community wants addressed. Then have preservice teachers work with students, teachers, and community members to create and carry out a plan for individual and collective action to address the bias and/or injustice. Finally, ask the preservice teachers to write a reflection on what actually happened when they took those actions to address the bias and/or injustice. This can then be a launching point to think about injustices inside of science education.

Teacher educators can also engage preservice teachers in learning to identify injustices in their own histories and then in their communities. Each year, our colleague Melissa Braaten begins her elementary science methods course with a "talking circle" and has preservice teachers discuss the harms from school science and science, as individuals. She asks them to think back to when they were in the age group they are currently teaching and to surface harms they experienced as learners: "What was it that made you feel less than? Or not heard?" This conversation then transitions to the present day, as she asks them to reflect: "Do we see this happening today? What would it look like to make things right? What did you need me [your science teacher educator] to do to make things right for you? How can we focus on repair rather than punishments and consequences?" Some students raise the issue of receiving poor grades in K–12 science classes and Melissa responds accordingly: "I will not give grades. I understand this has been harmful in the past, so let's find a new way to communicate with each other. Our goal is to come to an agreement about what's your best high-quality work." This initial talking circle sets the stage for ongoing conversations about school practices, such as group work routines, that need to be critiqued as damaging.

To keep this principle alive, preservice teachers and teacher educators can continually ask themselves questions such as: How do we problematize science and science teaching and restore harm done to students, teachers, and teacher educators such that they can reconstruct and redefine their relationship to science? How do we think restoratively about science classes? What institutional/structural changes need to change to support social and restorative justice in the school/learning environment?

In summary, the principles are most powerful when they are worked on simultaneously and applied as part of a larger transformation project across teacher education programs and the communities they intersect with. Our take is that all principles need to be an object of improvement, preferably at the same time. In the upcoming chapters, our colleagues further describe how they take up these critical equity principles to support Ambitious Science Teaching teacher educator pedagogies.

CHAPTER 4

Comparing Pedagogies to Support Core Practices in Two Secondary Methods Courses

MATTHEW KLOSER AND MARK WINDSCHITL

How is secondary science teacher education structured to prepare well-equipped beginners to enact core practices? In one Indiana location, pre-service teachers are prepared intensively—three hours each afternoon, four days per week during the summer months. At a Washington-based university, pre-service teachers are part of a master's program with two semesters of methods courses that meet for two hours twice a week. In Massachusetts, as part of a residency program, they gather in a local school where they go in and out of science classrooms and prepare teaching segments that they will try the following day with students. In New York City, a cadre of future science teachers is formed within the exhibit learning space of a world-renowned natural history museum. And among the distributed islands of Hawai'i, science methods courses occur mostly within the camera angle of a weekly video conference, with screen-sharing capabilities and virtual breakout rooms.

These examples are just five of hundreds of contexts across the nation, highlighting that teacher education programs are defined more by their variation in structure than by their similarities. This variation presents two significant challenges for those who educate science teachers. First, they must be prepared to model and support core practices in ways that work effectively across a broad range of program structures that serve an equally diverse range of K–12 school contexts. The second challenge lies in the absence of explicit training for potential science teacher educators. For many wishing to work with preservice teachers—including the two authors of this chapter—the extent of teacher education preparation included, at most, one seminar class in graduate school on theoretical perspectives of teacher education and, as mentioned in chapter 1, an apprenticeship of observation that sociologists of education so aptly describe as a major influence on classroom teachers' practice.[1] In this case, the apprenticeship occurred through our own time as classroom teachers and for one of us, through teaching assistantships in methods courses with veteran teacher educators.

This apprenticeship can be a positive experience in which a highly effective and experienced science teacher educator models practice, coplans with the graduate student, and makes explicit their decisions for shaping the experiences of preservice teachers. But even this apprenticeship has its limits. The teacher educator is focused centrally on the development of the preservice teacher, whereas developing the capacity of the future teacher educator is a secondary concern. Thus, what sociologists of education have written about developing one's teaching might well apply to developing teacher educators—"socialization into [teacher education] is largely *self-socialization*," placing the onus of learning on the graduate student through indirect means.[2]

Observing, coplanning, and participating in teaching methods courses is useful, especially with a helpful mentor, but time constraints and the opportunity for apprenticing in only one context with a specific structure limits the amount of metareflection and grappling with the problems of practice one can do. These problems can be recast as the central questions of teacher education: What works? For whom? And under what conditions?

This chapter is a window into our journey as teacher educators and the pedagogies used to prepare teachers to utilize core practices across two different sites. Specifically, we discuss four individual pedagogies that fall within the sets of pedagogies highlighted in chapter 1. The first two pedagogies, analyzing video of practice and modeling instruction, fall within the set of "Introducing and learning about core practice(s)" pedagogies. The next two pedagogies, coplanning with preservice teachers and facilitating rehearsals of practice, fall within the set of "Sheltered planning and practice" pedagogies. However, given the variety that defines science methods contexts, identifying surface features or guides to pedagogies would be relevant for only a few contexts. Rather, we frame and describe the pedagogies in terms of foundational principles. These foundational principles are characterized by:

1. Assumptions about aspects of the pedagogy that are valued as well as justifications for why they are valued.
2. Fundamental beliefs about the conditions under which preservice teachers or K–12 students learn in equitable ways and participate in authentic intellectual work.

Why focus on principles? They are the nonnegotiables of the pedagogy that will not vary across contexts. These principles make explicit that which is often only implicit to developing teacher educators—the reasons why particular pedagogies are important and how they shape equitable, coherent experiences for preservice teachers.

Before addressing the four focal pedagogies, we outline the three general principles that are foundational to all of the pedagogies we use. Each particular pedagogy then has secondary, pedagogy-specific principles described below. First, teacher educators make clear the concerns for equity and inclusive participation of students that are embedded within the teaching practices and in the teacher education pedagogies. That is, all pedagogies should provide an avenue for identifying potentially marginalized students and be enacted in ways that equitably address their needs and foster their participation. Second, pedagogies used in secondary science methods classes should focus on developing the practice of preservice teachers in the context of responsive teaching, student engagement, and sensemaking, not classroom management or procedural work. Third, teacher educators should make their pedagogical reasoning visible to preservice teachers (and when appropriate, future teacher educators) and model professionalism by opening up their own teaching practice to critique.

All of the following four pedagogies attend to these foundational principles and are accompanied by secondary, pedagogy-specific principles. We share these secondary principles and describe how they are reflected in the pedagogies themselves, highlighting within our two science methods courses their similarities and the variations that have resulted as responses to local structural differences and disciplined improvisation.[3]

PEDAGOGY 1: LEARNING FROM VIDEO

Pedagogy-specific principles

1. Video of teaching intellectually rigorous experiences in diverse classrooms provides the clearest insights into the kinds of responsive and equitable teaching that we value.
2. Preservice teachers should primarily focus their viewing on modes of and opportunities for participation by the students, rather than on teacher behaviors.
3. Instructors and preservice teachers explicitly discuss the frames of reference they are using to interpret action in the video.

At several points during our methods classes, preservice teachers begin deep dives into a set of core teaching practices that have been selected because of their flexibility across units and their ability to engage all students in intellectually rigorous work. Preservice teachers begin unpacking elements of teaching by reading informative texts, case studies, or transcripts that exemplify these practices and make clear how this instruction and the conditions for learning support students' disciplinary reasoning and their participation in equitable ways. In this "learning about" phase of our pedagogical cycle, classroom video is a powerful tool. Beginning educators are drawn to the dynamic exchanges among teachers and their students and the unpredictability of learning situations; our preservice teachers process what they see, hear, and feel in ways that shape their own aspirations for teaching. Analyzing video can be helpful at multiple points within the learning cycle; in this section, we focus on the use of other teachers' videos in whole-class settings.

Learning from video is not a straightforward process. Acts of noticing, interpreting, valuing, and reasoning all occur within frames that shape the elements to which we attend. Research on learning from video indicates that teachers notice some meaningful actions in the classroom while overlooking others, and teacher noticing is not merely a series of isolated events that occur consecutively in time or space. Rather, what a teacher observes in one moment influences, at least in part, what the teacher notices next.[4] All this makes the use of video with preservice teachers (or experienced educators) more challenging. Our first task is to shift the attention from teacher behaviors and classroom management toward analysis and interpretation of students' reasoning and participation. This shift is important to foster intellectually rigorous work grounded in students' thinking and to address biases and positionality that can lead to more culturally sustaining classroom experiences.

Choosing videos

Video that is recorded in culturally and linguistically diverse classrooms and that shows students talking and interacting has priority. For example, Mark uses video shot over the course of weeks in several classrooms because it becomes possible to "know" the teacher and the students better and to understand what norms and routines are in place in the classroom. It also helps to capture the teacher talking about their goals for the lesson before it is taught, and if possible, their commentary after the lesson. In many cases, the video is supplemented by student work samples to allow deeper insights into their thinking and their work together.

We seek out positive case examples of good teaching, but at the same time we feel that preservice teachers benefit from unexpected student responses and from students who are confused or frustrated, or have conflicts with peers. Watching a skilled educator does not preclude seeing episodes of missed opportunities, and these are beneficial for preservice teachers to unpack. In some cases, it may be less productive to watch students working together later in the school year, because the foundational labor of setting up the classroom norms, routines, scaffolding, and shared talk moves (aka, the "conversational infrastructure") and the building of relationships that lay the foundation for complex, collaborative engagement have been established by the teacher months earlier. This backstage work, with all its challenges, is important for preservice teachers to see.

Watching videos productively

Mark introduces the teacher and provides background on the students, school, and community as a way to contextualize and humanize the nature of the action the class will observe. He also has preservice teachers watch the video before they come to class, to accommodate those who want to spend more time processing or replaying what they are seeing.

The videos show parts of a full lesson because practices make better sense in relation to one another. For example, watching youth respond to a teacher who is circulating among groups to press for deeper reasoning makes more sense if we have just watched how that teacher framed the activity to the whole class and made explicit her expectations for collaborative work.

Video can be shown and analyzed with different purposes in mind. Videos might:

1. Provide a proof of concept (e.g., examine how a teacher recognizes students' ideas, out-of-school experiences, uses of language, or puzzlements as potential resources in a conversation)
2. Address concerns of preservice teachers (e.g., what if students seem "stuck" and unable to move forward in a lab activity?)
3. Show an example of . . . (e.g., check how modest levels of scaffolding allow emergent multilingual students to take the lead in group work)
4. Analyze a series of moves and responses by students (e.g., observe as students take turns, independent of the teacher, to offer evidence-based claims and critique arguments by peers)
5. Compare and contrast mini-cases (e.g., juxtapose the responses of students in two classrooms to different kinds of teacher framing before an activity)

Everyone is expected to contribute to deliberations about the videos. Mark has learned that it is important in his methods classroom for candidates to challenge one another with comments like, "Where did you see that happening?" or "I see this in a different way; it seems like an example of X, rather than Y." As these conversations unfold, however, we remind our preservice teachers that we are *not evaluating* the teacher or students. Miriam Sherin and Rosemary Russ, researchers in the area of video and teacher learning, provide examples to avoid: "I don't think that explanation is going to be very helpful" or "These kids are really good at participating."[5] They also identify a related interpretive frame in which observers *offer alternatives* to the actions that took place in the videos, such as, "Instead of letting students flounder with the vocabulary, she should have provided them with key terms earlier." Mark has named both the evaluation and alternative stances with his students as "repair talk." The label proves helpful for preservice teachers to use when making peers aware that their focus of analysis has veered away from students.

In Matt's class, preservice teachers watch segments of video four or five times during the course. For some videos, he has found it helpful for preservice teachers to use an initial noticing frame, which helps them recognize elements of the practice that will be decomposed during discussion of the video. In other instances, they use an explicit framework, such as a "framework for facilitating a whole-class sensemaking discussion" that Matt developed to help organize and guide their noticing and analysis.

Video analysis and reflection

While we focus on the interactions among teachers and students to accomplish particular goals, we also apply a critical lens to the lesson by asking: Who is being served well by the predominant norms for talk? Do students have a chance to develop agency as knowers? Who is being heard and whose ideas are being discussed? What does this kind of instruction prepare students for? Who is on the margins of this activity? While we often show equitable and inclusive practices in the video, we want our preservice teachers to know that questions about power and privilege have to be asked of any classroom episode, especially their own. In Matt's class, teacher candidates create a graphic organizer for reflecting on the task in which students engage, the types and uses of tools—physical, conceptual, and social—for the teacher and the student, and the nature and forms of talk. As the teachers generally do not watch the video ahead of class, this organizing structure facilitates more focused and complex noticings.

At other points in the learning cycle, preservice teachers can watch their own video—for example, after they have rehearsed a set of practices or tried out lessons in a classroom. Once they understand how complex responsive teaching is in live settings, they can use video to assess the differences between what they had planned for and what actually unfolded.

PEDAGOGY 2: MODELING TEACHING PRACTICE

Pedagogy-specific principles

1. The instructor models forms of teaching that are responsive to students' ideas, puzzlements, and everyday experiences.
2. The instructor represents risk-taking and vulnerability, opening up their practice to critique and questions by preservice teachers to support their learning.
3. Preservice teachers learn by participating as students in challenging disciplinary work and also by doing guided reflection on what aspects of instruction influence their interactions with others or their thinking.

Modeling, or demonstrating a set of teaching practices, is more than just "running through an activity" with preservice teachers. It is a mutually constructed set of interactions between someone playing the role of K–12 teacher and other adults playing the roles of young learners. These interactions are both structured and improvised in order to be responsive to students' questions or unexpected pathways of engagement. Modeling allows preservice teachers to feel what it is like, in the role of students, to respond to prompts, to have their conversations scaffolded, to use a tool provided by a teacher, to feel a growing sense of competence or frustration with a science activity, and perhaps to feel what it is like to have their ideas treated as resources for everyone's reasoning.

For the instructor, it is a chance to model not only the act of teaching, but also the vulnerability inherent to ambitious and unscripted pedagogy. In our own episodes of modeling, we have felt our adrenaline flowing as our "students" take conversations in directions we hadn't anticipated, or give us shocked looks as we inadvertently dismiss a classmate's idea. These situations are opportunities to show humility about our own teaching and acknowledge that uncertainty in the midst of a performance is normal for the kinds of teaching we aspire to and integral to breakthroughs in professional learning.

While representing practice is one of several pedagogies that can be used to introduce practices to preservice teachers, we do not view it as an optional part of our repertoire. If we, as teacher educators, cannot simulate ambitious instruction even in the protected environment of a methods class, then we do not really understand the practice, nor do our preservice teachers get chances to understand what the practice feels like to students. We have an obligation to open ourselves up to the same critiques that our preservice teachers will experience.

What we represent and how

In one episode of modeling teaching, Mark typically enacts two or three practices that build on one another during a lesson. For example, in a lesson on eliciting ideas, he starts by framing the upcoming activity, describing how the topic might connect with students' interests, and then reminding them about the range of ways they could share their initial reasoning about a phenomenon. Mark then probes students' observations and mini-theories through structured conversations. Finally, the preservice teachers, participating as students, make their thinking visible through draft scientific models or public lists of ideas.

Mark models these practices because they require responsiveness to students' thinking and the use of their ideas and out-of-school experiences as resources to engage in rich disciplinary work. Furthermore, this type of modeled teaching segment invites students to reason collaboratively about ideas that change over time, and to demonstrate equity and inclusiveness in all interactions.

Despite the intentionality in choosing what is represented, the enactment varies dramatically each year. This is, in part, because we learn from what our preservice teachers take up and experience in their own rehearsals and clinical experiences. If Mark sees in their later classroom videos that they struggle to include all students in small-group conversations, then the following year he makes a special effort to emphasize this aspect when representing and discussing practice. These iterative cycles of learning with and from students are important because they are foundational aspects of Ambitious Science Teaching; novices can be made explicitly aware that changes occurred based on interactions with previous cohorts, setting the stage for preservice teachers' own ongoing adaptations in their classrooms.

While modeling teaching practice, Mark uses two strategies for enhancing what novices learn from the enactment—*getting paused*, and *stepping out of his role* to be explicit about pedagogical decisions. Getting paused means that a preservice teacher signals Mark to stop so they can comment on a particular move that was made or to "rewind and replay" the last few moments using an alternative strategy. Because preservice teachers are initially hesitant to pause their professor, Mark asks his teaching assistant to do this several times during the first modeling segment. Everyone can witness what is worth stopping the action for and how these spontaneous conversations highlight problems of practice.

Besides adding value to representations of practice, pausing gives preservice teachers a sense of control over the collective work and what they can learn. It also normalizes the stopping of enactments to question or suggest changes. In essence, Mark is also representing how to respond calmly to requests for pausing, to entertain a brief conversation, ask for help, and then collect his thoughts and continue. This prepares preservice teachers to be paused when they do their own rehearsals or to call "time out" themselves—a kind of self-pause—to get input from the teacher educator or peers about how to move forward.

Mark also steps out of his role to be explicit about why he used a particular talk strategy, made an adaptation to an activity after it was under way, or decided to offer scaffolding to support student thinking. He might stop the action during a whole-class discussion and say:

> Let me step out of my role here; Josh just had a thirty-second turn at talk, explaining why this ecosystem got disrupted. In a classroom, some students would have trouble tracking the two or three big ideas he embedded in that narrative. I'm just going to briefly revoice what he said without overwriting his ideas with my own, and emphasize those key concepts.

In cases like this, the benefits of having a shared language about talk become clear. The class is all familiar with revoicing, elevating an idea, suggesting crosstalk among students, wait time, follow-ups, pressing, and probing. All of these are tools that the teacher can use to accomplish communicative knowledge-building goals with students, but in the context of our methods class we can also use these terms for more precise analysis of what is being said, by whom, and how.

After preservice teachers participate in these enactments, they know they won't be replicating what Mark does because they'll be planning for a rehearsal that focuses on different science topics, and their enactment will have to adapt to what their peers say or do during a rehearsal. What will guide them are the supportive contours of the teaching practices, the commonly understood goals for this kind of instruction, an adaptable set of tools and scaffolds, and the principles that underlie any enactment of these practices. In Matt's class, pauses are more infrequent during the modeled teaching segments, but preservice teachers are asked to track their own noticings and questions that are surfaced during a formal debrief. During the debrief, Matt shows the preservice teachers the goals for the learning segment and how what was enacted correlated to an original lesson plan, illustrating how that plan had to change in light of student ideas and thinking. The debrief then helps the preservice teachers pull together big ideas about teaching, ultimately helping them shape the planning for their own attempts at working with young learners.

PEDAGOGY 3: COPLANNING BETWEEN METHODS INSTRUCTORS AND PRESERVICE TEACHERS

Pedagogy-specific principles

1. Coplanning raises critical questions about how the tasks, tools, and opportunities for talk foster equitable participation and thinking among all students.

2. Coplanning makes explicit the often-implicit justification for how and why learning segments address various student learning needs.

3. Coplanning focuses on the development of big science ideas and avoids designs for procedural or scripted instruction.

All teachers plan. But not all teachers plan for big ideas in science or for students making connections between phenomena and their causal explanations. Planning for meaningful work in science classrooms is a complex task, and coplanning is an opportunity for sheltered practice with the teacher educator, helping highlight what might often be implicit.[6] Like all pedagogies, coplanning can take various forms, but in all versions, it is a supportive opportunity to identify an ambitious, student-centered learning goal and the means by which students can do rigorous intellectual work.

Coplanning for course coherence and big ideas

The goals of coplanning look different at different granularities. At the course level, preservice teachers need help moving from developing discrete units toward more coherent stories of physics, chemistry, biology, or earth science that play out over eight to nine units. Matt has learned that the first coplanning experience is most helpful when taking a course-wide view; preservice teachers are given a set of twelve to fifteen nondescript topics like "genetics," "one-dimensional motion," or "weather and climate," and are then asked to organize and prioritize concepts. Once table groups finish organizing these topics, Matt pauses the whole class and works with one topic from one group. In a recent class, Matt used the topic "ecosystems" to push preservice teachers to think about what this vague title might encompass. Drawing from their prior content knowledge as well as the standards, he asked preservice teachers to write three to five sentences that capture the big ideas that one would hope their students would still understand about ecosystems thirty years after taking the course. This level of coplanning intends to provide preservice teachers a sense of coherence within their courses.

Coplanning for rehearsals

Coplanning for rehearsals—approximations of teaching practice that are described below—is another common occurrence. As preservice teachers prepare to teach their peers, coplanning with other preservice teachers and the teacher educator is a central avenue for feedback. Rehearsing the facilitation of a sensemaking discussion around data, Matt has to narrow the scope of planning due to the large number of preservice teachers who need to rehearse. Rather than plan and run an investigation that provides data, preservice teachers are given one of seven different data representations tied to a core disciplinary idea. Although the "students" in the rehearsals do not collect the data themselves, they are shown the investigation that brought about this data and given the context that produced it. The representations include data that comes from investigations such as the potential and kinetic energy graphs of a cart going down a track or authentic data from a study on the effects of an oil spill on bird populations on the Portuguese coastline.

While coplanning, Matt pushes preservice teachers to think about the task, tools, and talk that can be leveraged to help students make sense of the data. Matt moves among small-group tables to help preservice teachers think about how they can make connections between the data and the material activity. He raises questions about where young learners can do the thinking but might need help organizing their thoughts or comprehending the data. He asks groups about what tools—conceptual or physical—might ensure all students have access to participation in the discussion. Finally, Matt and the preservice teachers work together on the central questions that will guide the discussion, thinking about possible student ideas and what follow-up prompts might be helpful based on the projected range of what students perceive.

Mark's coplanning follows the same three guiding principles as Matt's, but his preservice teachers have slightly different experiences. Mark introduces his teacher candidates to the standards and the design of units that engage young learners in sustained intellectual work on complex science problems. When he starts his first pedagogical cycle—introducing a set of core practices, modeling them, and decomposing them—he asks his preservice teachers to select one of four topics related to climate change (ecosystem disruption, combustion of fossil fuels, greenhouse effect, ocean as heat sink). Preservice teachers work together and with Mark's guidance, develop an individual lesson that incorporates the teaching practices featured in that cycle. They also sketch out the larger unit within which this lesson would be taught. Because Mark's course does three full pedagogical cycles during the quarter, his preservice teachers develop and rehearse three very different kinds of lessons. Later in the quarter, they support each other in revising their draft units.

Coplanning is a fluid pedagogy—one that is defined by the nature of what is being planned, by the structure of the science methods course, and by the proximity of the planning to actual instruction. But as detailed in the examples above, it always addresses how students can enter into the content. This requires making explicit—often by making one's own planning explicit first—the reasons for how and why segments of instruction are more or less effective at using and working with students' ideas.

PEDAGOGY 4: REHEARSING TEACHING PRACTICE

Pedagogy-specific principles

1. Opportunities to learn from rehearsals are optimized when they require preservice teachers to address students' ideas, the disciplinary content, the relationship with students, and their engagement in science practices simultaneously.
2. Feedback and the interrogation of practice by the community are most effective when occurring during pauses in the enactment.
3. Rehearsals are a collective learning experience in which the structure should elevate the voices of the preservice teachers and the teacher educator during pauses and debriefs.

Why rehearsals?

Most early career teachers live by the adage, "We learn from our mistakes." Teaching requires scores of decisions about content, student thinking, and instructional moves, and while not all of these decisions have the desired effect, when reflected upon, they can lead to important professional growth. Mistakes and ineffective teaching cannot be eliminated, but when teaching is viewed as an act of social justice, teacher education must do its best to reduce instruction that is unresponsive to students' ideas or inattentive to who is participating, especially for the most vulnerable young people. Therefore, having preservice teachers "try out" teaching practices for the first time on real students should be minimized.

Running rehearsals

Teaching rehearsals are one approximation of practice that provide preservice teachers opportunities to try particular aspects of teaching within the shelter of a trusting community.[7] Rehearsals, like instructor modeling, distinguish themselves from other forms of enactments by the intentional pauses within the teaching segment that allow not only the preservice teacher who is teaching, but the entire classroom community, to work on how the pedagogy unfolds.[8] While they can take many forms, rehearsals in our methods courses generally follow a co-planning activity. Preservice teachers are assigned roles as students and provided with expectations for how they should act. Similar to modeled segments of teaching, pauses, collective thinking, and rewinds are used to work on teaching in the moment. As human working memory can be unreliable when debriefing at the end of a thirty-minute segment, using pauses in the midst of practice reduces participants' need to rely on inaccurate memories while providing opportunities to not only discuss, but also try out different pedagogical options. For example, when a preservice teacher poses a closed-ended question that limits the intellectual demand on students, the teacher educator can pause, ask all the members of the class what type of question might result in more reasoning from students, ask the focal preservice teacher to rewind and offer this new question, and then see what, if anything, changes in the discussion. Depending on the time available, the rehearsal ends with a whole-group debrief and if videotaped, an opportunity for ongoing reflection for the preservice teacher.

All preservice teachers and the teacher educator have an important voice in analyzing practice and providing feedback to the rehearsing teacher. In some ways, this idea is an extension of the first general principle that crosses all teacher education pedagogies—concerns for equity and inclusivity are made explicit by the teacher educator. Empowering the voices of all preservice teachers during rehearsals signals that even less experienced teachers provide an important voice to one's instructional growth. This principle also suggests that not only should all members of the community feel their voice is important, but each learning opportunity for one preservice teacher is a necessary learning opportunity for the whole community. All preservice teachers and the teacher educator learn together, even when acting as "students."

Rehearsals in context

The final rehearsal in Matt's first-year science methods course focuses on the complex practice of facilitating a sensemaking discussion of a core idea based on data. Given the number of students and the timing of the course, it is not possible for each preservice teacher to plan a unique investigation from which "students" could collect data. Therefore, groups of three or four students are given one of seven visual representations that contain data related to a core disciplinary idea in earth science, physics, chemistry, or biology. For example, preservice teachers teaching earth science are given a world map containing icons of different fossil types present on different continents that could be used to help students construct ideas that landmasses were once positioned very differently—setting the stage for developing an explanation of plate tectonics as a major causal force for the structure of earth's landforms. The decision to provide the data reflects constraints in Matt's own context while still adhering to the three foundational principles of the pedagogy outlined above. Furthermore, the curricular piece is provided so that preservice teachers can focus less on constructing an appropriate lesson plan objective and more on developing central questions for the discussion.

Each rehearsal lasts twenty minutes, with the preservice teacher providing two minutes of context and instructions for the "students"—usually four or five peers, with three rehearsals happening simultaneously in different rooms, facilitated by the teacher educator and two clinical faculty members. The preservice teacher is reminded that they can pause or might be paused by the teacher educator. On average, a twenty-minute rehearsal involves three to five pauses and is videotaped for later reflection.

The following excerpt follows the "teacher" (T), Laura; five peers who are students (S); and Matt as the teacher educator (TE). Laura is using the above described map in order to get students to begin making data-supported claims about the dynamic nature of Earth's landmasses. Italicized text is used to indicate when preservice teachers are acting in the roles of teacher and students.

LAURA (T): *So, where do you see fossils of Kannemyerid?*

THOMAS (S): *On Africa, South America, and, um, should I keep going?*

LAURA (T): *Uh-huh. Sure.*

THOMAS (S): *North America and Asia.*

LAURA (T): *Okay. So, if they are here [points to the Northwest coast of Africa], how might they get here [points to the northeastern coast of South America]?*

KATIE (S): *They could swim?*

PATRICK (S): *Or maybe they evolved there?*

LAURA (T): *Okay. Not quite what I was looking for . . .*

MATT (TE): Let's pause here for a second. [The teacher candidates and Laura look to Matt.] The last comment, when you said, "Not quite what I was looking for . . ."—I wonder if this made students, um, think that this was like a guessing game. They had a couple of ideas, but I wonder if they thought that they just had to figure out what you were thinking. This next question is for anyone, not just Laura—what had "students" [does air quotes] offered and what might Laura do to work with those ideas?

ALBERTO: Well, there are like, two things going on with the swimming and the evolution, but that's not what you really want to talk about.

PATRICK: Thinking as a student, I actually was kind of disappointed when you said that it wasn't what you were looking for. But I know you don't have, um, time for all of them, like for talking about all of the students' ideas.

KATIE: I wonder if you might just get a couple of more ideas from students and then ask another question. I'm not really sure what to do though, because you don't want to, you know, like take a ton of time for each idea.

LAURA: I want to get them to a point where I tell them that these parts of land were once together. They don't know that. Should I just tell them? But then if they know that, then I can get them to the bigger point, maybe, that they could, like think about, that in order for them [the dinosaurs] to now be apart, these things [pointing to landmasses] had to, like, move or something.

[Silence for about ten seconds]

KATIE: Well, maybe, if you don't want them guessing and stuff, about the land-masses being together, then just tell them right away and then ask something like, "What do you think would have to be going on if we know that these dinosaurs can't swim and couldn't evolve into the same species in four different places?" Then maybe it would get students thinking that the land had to be connected or something, and then you could, like, start asking about what would have to happen to go from connected land to an ocean between them?

ALBERTO: Yeah, I was thinking something like that.

MATT (TE): All right. So let's rewind to where you were pointing at the different parts of the [map]. And, um, ask, like, something like Katie just said, and see what students think about and talk about then. Maybe see if that moves you closer to getting them to making connections with the data and the potential causal story here. You know with the plates, [these are] things that they can't observe, but, like, maybe they could use the map with evidence to pose one plausible reason.

This rehearsal excerpt highlights how one structure can address this pedagogy's foundational principles. By pausing Laura in the moment, the entire class had the opportunity to think specifically about the question posed, how it was posed, and what student thinking resulted from asking this question. This differed from comments that often happen during the debrief that end up vague and often evaluative, such as one that occurred during the debrief of this rehearsal when one candidate commented, "That was really good when you asked the question about the fossils on two continents because students really had to think." By posing the question to the entire group, the teacher educator also indicated that it was important for the class to come to some collective understanding that did not reside solely with the teacher educator, and that there are multiple variations that could foster student learning.

In Mark's methods class, similar rehearsals are thirty minutes long and involve connected sets of practices like (1) framing the intellectual work expected of students, (2) engaging students in small-group work while pressing their thinking, and (3) facilitating a whole-class sensemaking discussion. If this sounds like a lot for each session, and that it could be a bit chaotic, you would be right. The "students" however, know ahead of time what the small-group activity will be, so no instructions are given, meaning that after the teacher frames the discussion, they dive right in. The sensemaking discussion rarely comes to a satisfying conclusion, but the preservice teachers feel a sense of accomplishment in inviting and sustaining collaborative reasoning. These are the trade-offs when high expectations for novices' attempts at responsive practice meet the constraints of higher education.

CONCLUSION

Preparing teachers for the range of contexts in which they might serve is a complex task. If we can identify effective teacher education pedagogies that can be used coherently and flexibly across various structures of methods courses, we can increase the chances that preservice teachers have opportunities to attend to the tenets of ambitious forms of science teaching. As the descriptions and examples from our classrooms suggest, these pedagogies require teacher educators to take risks and make their teaching open to public critique. And while the pedagogies look different across different classrooms, adhering to foundational pedagogy-specific principles helps foster equitable opportunities to learn for preservice teachers.

CHAPTER 5

Lessons and Challenges from Three Years of Preservice Teachers' "Macroteaching"

AMELIA WENK GOTWALS, BRIAN HANCOCK,
AND DAVID STROUPE

As noted throughout this book, Ambitious Science Teaching (AST) can serve as a pedagogical framework for teachers to support students learning through participation in science and engineering practices. Given the goals of AST, supporting preservice teachers as they learn to enact such complex instruction is crucial. One difficulty teacher educators face as they plan methods courses is that preservice teachers rarely encounter instruction that resembles AST during their time as students—either in K–12 or postsecondary contexts. Therefore, preservice teachers' default visions of teaching often resemble the instruction they experienced as students.[1] Teacher educators, then, have two important tasks: (1) disrupting preservice teachers' initial visions of science teaching, and (2) providing preservice teachers with opportunities to learn complex, rigorous, and equitable instruction. In our courses, such instruction is AST.[2]

We propose that using a practice-based framing of teacher education can provide preservice teachers with opportunities to approximate the core practices of AST in their methods courses, thereby providing a simultaneously disruptive and supportive context in which to learn.[3] Thus, methods courses can provide the type of community in which preservice teachers learn to make pedagogical decisions that arise during moment-to-moment interactions with students in classrooms.[4]

MACROTEACHING

In this chapter, we describe a practice-based learning opportunity we codesigned with preservice teachers in our methods course to support their learning of AST. Dubbed "macroteaching" by a participant, this learning opportunity is an extended pedagogical rehearsal in which preservice teachers take on the role of "lead teachers" for their peers.[5] Macroteaching arose in response to the preservice teachers' critiques that, while their initial microteaching rehearsals were helpful (see below for more details), such experiences did not feel "authentic enough" in terms of approximating the intellectual rigor and equitable learning opportunities they envisioned possible through AST.

Macroteaching has thus far occurred at Michigan State University, where preservice teachers participate in a five-year teacher preparation program. During the program, preservice teachers major (and minor) in a science discipline (e.g., biology) while also fulfilling requirements for a teaching certificate. During the preservice teachers' senior year of college, they participate in two secondary science methods courses—one in the fall semester and one in the spring semester. During the fall semester, preservice teachers learn about the Framework for K–12 Science Education and AST through readings, watching videos, and engaging in representations of the core practices by course instructors.[6] In addition, preservice teachers approximate core practices during microteaching opportunities within the methods course.

Macroteaching was codeveloped by David Stroupe, Amelia Wenk Gotwals, and a cohort of preservice teachers, based on the their three primary critiques of microteaching: (1) the microteaching episodes were too short (i.e., twenty minutes, as opposed to longer lessons that are typical in many secondary classrooms); (2) too much time elapsed between microteaching opportunities (about two weeks); and (3) preservice teachers taught three or four peers acting as students, which limited the number of student ideas they could elicit and use to inform and adapt their teaching. Given these critiques of microteaching raised by the preservice teachers, we decided to represent a feature of AST—adapting our instruction as teacher educators based on students' expressed and emerging needs.

Using the preservice teachers' critiques of microteaching as evidence of their emerging needs, we codeveloped macroteaching during the spring semester methods course to allow for a peer-teaching opportunity that might better resemble the daily work of teaching in secondary science classrooms. Macroteaching involved having groups of preservice teachers plan, teach, and reflect on eleven to twelve consecutive hours of instruction to their peers (approximating a complete unit of instruction) during methods class. In the initial macroteaching experience, we learned with and from the preservice teachers as they helped us to codesign the experience.

Through analysis of video and preservice teachers' plans, reflections, and interviews, we found that preservice teachers' understanding of professional work changed in two fundamental ways through the macroteaching experience. First, they became more comfortable navigating uncertainty when they actively sought and valued students' ideas. For example, they shifted to seeing students' "curve ball" ideas as potentially informative and productive instructional resources rather than "misconceptions" to fix. Second, the preservice teachers noted that by having to serve as instructors *and* students during macroteaching, they experienced the discursive and pedagogical moves found in AST from multiple perspectives. Such experiences helped the preservice teachers begin to see how students could work as knowledge builders in the classroom as they leveraged tools to organize and use knowledge to identify and solve complex problems.

Since the initial cohort in 2016, we have enacted macroteaching with two additional cohorts of preservice teachers. In this chapter we share lessons learned during the three cohorts of preservice teachers who have "macrotaught" in the second semester of their senior year methods course. Specifically, in the following sections, we highlight how this extended pedagogical rehearsal affords both the preservice teachers and us, as teacher educators, the opportunity to promote rigorous and equitable learning communities. Next, we show how macroteaching has served as an opportunity for preservice teachers to try out a range of pedagogical moves in a safe environment while situating specific practices and tools in a larger, coherent picture of professional work. Finally, we highlight how macroteaching provided us with opportunities to gather evidence of preservice teachers' learning, and to provide immediate feedback to preservice teachers about their teaching.

MACROTEACHING AS A COLEARNING OPPORTUNITY

Over three years of enacting macroteaching with preservice teachers, we have noticed four emerging features of a practice-based methods class that offer opportunities for us and our students to grow as teachers.

Opportunity 1: Teacher educators colearn with preservice teachers

During the first year of macroteaching, we were responsible for the original framing and planning of the pedagogical experience. As we explained the plan for extended instruction to the preservice teachers, we were prepared to address questions and confusion, and to adapt the experience as their needs emerged. We did not anticipate, however, that preservice teachers would utilize macroteaching to create immediate learning opportunities that we would carry forward into future iterations of the pedagogical experience in subsequent methods courses.

During the first few lessons, we realized that, rather than impose too much structure onto macroteaching, we needed to codevelop the learning experience over time with the preservice teachers. By codevelop, we mean that the preservice teachers and methods instructors had opportunities, both in the moment during instruction and in assignments, to articulate emerging and shifting learning needs and to advocate for solutions. Sometimes the solutions emerged during instructional episodes during methods class, and other times we reflected on the proposed solutions and made modifications in the following class.

Over the course of three years, five primary learning opportunities have been codeveloped with the preservice teachers during macroteaching. The preservice teachers and methods instructors agreed on a name for each learning opportunity, thus establishing a shared language and resource for future cohorts of preservice teachers. These are the five colearning opportunities:

- *In-the-moment consultations.* The teaching team gathers to discuss their immediate and upcoming instructional decisions, often done during transitions between episodes of instruction.
- *"Time-out/time-in."* Preservice teachers pause their instruction—calling "time-out"—in order to engage in immediate reflection about a recent

event. After a public conversation and consultation, an individual or team restarts the pedagogical action by calling "time-in."

- *"Rewind."* At times, following the time-out/time-in or instructional coaching, a teaching team member requests an immediate opportunity to retry an interactive episode. Often, preservice teachers request a "rewind" based on their colleagues' advice.

- *Question-and-answer session at end of each lesson.* During the final five to ten minutes of each lesson taught by the teaching team, the preservice teacher "students" and instructors ask the teaching team to decompose pedagogical decisions they made during class, and inquire about upcoming lessons. This process varies by year, with the latest cohort writing "see-think-wonder" feedback statements to deliver to the teaching team after each lesson.

- *Teaching team debrief at the end of the unit.* Similar to lesson debriefs, these conversations occur at the end of the unit. The preservice teacher "students" and instructors ask the teaching team to decompose pedagogical decisions made during the unit, and to reflect on instructional opportunities they might shift in the future.

Opportunity 2: Preservice teachers learn together

A second feature of macroteaching is that over the course of the semester, the preservice teachers learn with and from each other by serving as both instructors and students in the methods course. For example, preservice teachers had the opportunity to experience AST from the vantage point of teachers designing a unit as well as from the perspective of students tasked with constructing and revising an evidence-based explanation for a puzzling phenomenon. Seeing AST through a student perspective became important to the preservice teachers as they considered the high level of intellectual rigor and equitable participatory moves they planned into their own units of instruction. For example, during the first year of macroteaching, a preservice teacher named Jack reflected, "As students, we've got to feel what it's like to get pressed and to work on each other's ideas over time." As Jack noted, and since most of the preservice teachers had not participated in science classrooms where their ideas guided instruction, they found value participating in a unit that their future students would experience.

While we were pleased that the preservice teachers eagerly took on the role of students, we initially worried that they would somehow disrupt macroteaching by knowing "too much," asking purposefully esoteric questions, or feigning ig-norance of a subject. However, we found that given the range of science majors in the class (i.e., biology, chemistry, earth science, physical science), many pre-service teachers did not deeply understand core ideas in various units. Thus, students generated hypotheses about phenomena that often included partial un-derstandings or ideas similar to those that their future secondary students might state. In addition, students could ask authentic questions, share emerging ideas, and learn about science from their peers. For example, during a macroteaching unit on the rock cycle, a biology major, Emma, said, "I like when I don't have to pretend I'm a student and can just ask questions." The opportunity for preservice teachers to experience the longitudinal role of a student in a safe environment was important for their developing vision of AST.

Opportunity 3: Making space for preservice teachers to encounter uncertainty

As our preservice teachers noted in their critiques of microteaching, learning pedagogy that hinges on public discourse and building relationships with stu-dents, such as AST, requires connected opportunities to approximate practices and adapt instruction based on what students say and do.[7] The development of pedagogical judgment to respond to unanticipated features of instruction, such as student thinking and actions, is a difficult aspect of ambitious teaching to learn.[8] Therefore, preservice teachers need opportunities to "notice" and use students' ideas as resources to make instructional adaptations.[9] In our methods class, we wanted preservice teachers to rehearse exercising pedagogical judgment in the face of uncertainty. Therefore, macroteaching became a learning opportunity for preservice teachers to encounter unanticipated talk and actions, to make purpose-ful pedagogical decisions based on the emergent student discourse, and to move forward with instruction based on the decisions they made in the moment.

The preservice teachers reported the value of encountering and making pedagogical judgments about students' emerging science ideas. As opposed to microteaching, during macroteaching, preservice teachers had time and oppor-tunities to elicit students' ideas and use them in productive ways both in the moment and day to day. Yet, such work remains difficult—across three years of macroteaching, we have found that most preservice teachers note the complexity of working with students' emerging ideas. As one participant noted, "Throughout the entire unit, there were multiple instances where I felt very uncomfortable, but I am happy to have had the opportunity to feel flustered for the first time with the help of my peers and professors rather than on my own in front of a room full of high schoolers." This preservice teacher expressed a shared sentiment of her peers: they wanted to initially experience the uncertainty inherent in AST *during* methods class, rather than in front of secondary science students. Macroteaching, then, allowed the preservice teachers to make decisions about uncertain moments in a safe and collegial learning environment.

Opportunity 4: Learning as teacher educators

The fourth feature of macroteaching that has emerged over the years consists of three continual opportunities to learn as teacher educators. First, we initially viewed macroteaching from our perspective as researchers of teacher education and AST. We assumed that we could readily identify and name difficult problems that might arise for preservice teachers when learning and approximating AST. While this was true in some sense, we quickly learned that the preservice teachers could name and identify tensions, problems, and conundrums that we could not imagine. As the preservice teachers began to exercise agency in their teaching and in creating novel learning opportunities (such as time-out/time-in), we realized that our role as facilitators of macroteaching needed to shift. Rather than simply prescribe learning opportunities, we realized that we could embody a foundational principle of AST—that students and teachers can codesign the learning opportunities needed to advance a community's work.

Second, macroteaching quickly forced us to confront and publicly advance our understanding of AST beyond naming and demonstrating core teaching practices for the preservice teachers. Instead, we had to dive deeply inside the core practices, understanding the framework from the teachers' perspective and reifying the professional expectations inherent in AST as methods instructors. We extolled and represented the value of pedagogical experimentation, noting for the preservice teachers when, how, and why we made adaptations to the methods course based on our developing understanding of students' emergent needs. We also noted when we became stuck pedagogically, and when we used the preservice teachers' ideas to shape macroteaching and the methods course. This public elevation of our thinking blurred lines between typical boundaries of pedagogical "expert" and "novice," thus representing a foundational principle of AST for the next generation of practitioners.

Third, over three years we have worked with multiple graduate students as coinstructors in the practice-based methods class. Similar to the preservice teachers, the graduate student coinstructors have identified features of AST and methods class pedagogy that might have remained unnoticed by the faculty instructors. For example, in the second year of macroteaching, Brian Hancock cotaught the methods class with David Stroupe. Over the course of two macro-teaching units (a chemistry group focused on fireworks and a physics group focused on a car collision), Brian noticed that the two teaching teams had difficulty in selecting the primary features of the phenomena to focus on as foundational for making sense of the complex event. Brian helped the teaching teams learn how to prioritize the main conceptual ideas of a phenomenon, while still attending to the goals of AST. In addition to helping faculty see new problems to solve, the graduate student coinstructors also learn how to plan, enact, and reflect on a methods course that focuses on core practices and AST. Thus, macro-teaching serves as an opportunity for the next generation of teacher educators to learn with faculty and preservice teachers about their future work.

TRYING OUT PEDAGOGICAL MOVES AND BUILDING RELATIONSHIPS

As noted previously, a crucial part of any methods course is for preservice teachers to learn how to build and sustain relationships with students. While core practices and tools from AST are effective in supporting both student and teacher learning, merely treating the practices as separate from the humanizing features of teaching can result in a "sterilization" of the practices in which teachers falsely separate instructional moves from relational work.[10]

We argue that a key feature of learning to skillfully enact core practices *is* building relationships with students, and in our experience, engaging preservice teachers in macroteaching represents one important opportunity to support their developing understanding of and ability to build rich relationships. In this section, we provide one example of macroteaching in which preservice teachers experienced—and successfully navigated—an instructional moment of uncertainty that highlighted how important relational work can be as learners engage in complex science practices.

Whale-nami

Our example began with an instructional team of preservice teachers from the second macroteaching cohort leading an earth science unit. The instructional team drew on the phenomenon of a graveyard of whales that was uncovered during a highway expansion project on the west coast of Chile. This area—dubbed Cerro Ballena ("whale hill")—is located a few kilometers inland away from the Pacific Ocean. The instructional team's driving question for this unit was, How did these whale bones end up in the desert, far from the ocean?

During the introduction of the unit to their peers/students, the instructional team hoped to elicit their students' ideas about the phenomenon (including ideas about plate tectonics and weather-related processes) and leverage those in subsequent lessons. After presenting the phenomenon to the class, the instructional team asked their students about relevant features of the phenomenon, and eventually asked them to think about how and why the well-preserved and intact whale bones could end up so far away from the ocean coastline. Each group of students was then instructed to develop an initial model, illustrating the before, during, and after of the whale phenomenon.

Three of the four student groups shared models that aligned with the instructional team's expectations of what students might construct as an initial explanation. These models included features such as tidal patterns and sedimentation (over a time period of thousands of years). One student group, however, provided an explanation that the instructional team did not expect. This group proposed that a tsunami lifted the whales out of the water and carried them to the Chilean desert, where they ultimately came to rest and were slowly buried over time. "Whale-nami," as this model became known, provided a moment for the instructional team to recognize what Paul Cobb might describe as a "point of departure" from instruction that prioritizes the memorization and recitation of information.[11] According to Cobb, a point of departure is a moment during instruction in which the teacher makes a consequential pedagogical decision that results in students having more intellectually rigorous and equitable opportunities than they might in classrooms that privilege knowledge accumulation and recitation.

In the context of whale-nami, the instructional team faced a decision about what to do with the students' model, and more importantly, how to treat the students' ideas. The decision was not taken lightly. The instructional team could have perceived whale-nami as "off-topic" or "silly." As the whale-nami group joined others in sharing their initial model with the class, the instructional team could have laughed aloud (with some students in the class) and ultimately discounted the value that the students' ideas added to the conversation. However, given the focus on building and sustaining relationships with students, the instructional team chose to see "whale-nami" as legitimate, and to publicly acknowledge the value of the ideas that led to the initial model. Further, whale-nami continued to be discussed throughout the unit as a possible explanation. By placing value on the students' unexpected ideas about the whale phenomenon throughout the unit, the instructional team, and the preservice teachers serving as students, were able to experience the importance of valuing and working on ideas over time in a setting that emphasized relational work as crucial to success.

During the three years of macroteaching with different groups of preservice teachers, we have found these "points of departure" moments to be monumentally important for building a community in which preservice teachers feel comfortable to share ideas and develop a vision of teaching as building and sustaining relationships with students. In the whale-nami example, the preservice teachers leading the macroteaching unit ultimately chose to listen and learn with their students rather than discount their ideas during the moment of instruction. However, this critical decision was not made without pause. Instead, during this pivotal moment (when the whale-nami group shared their initial model with the class) the preservice teachers and teacher educators, alike, collaboratively worked together during an instructional "time-out" to discuss potential next moves with the entire class.

We highlight three important affordances of macroteaching as a rehearsal for trying out pedagogical moves and building relationships. First, the preservice teachers leading whale-nami had a community of support—the teacher educators and their peers—to try out AST in a safe environment. Knowing the importance of placing value on student ideas, but unsure of how to proceed, the preservice teachers utilized an opportunity to encounter the uncertainty that is inevitable in classrooms, and to have a community of colleagues who understand the complexity of the upcoming decision assist them in considering next pedagogical steps. By thinking out loud about whale-nami as a community in the context of macroteaching, all of the preservice teachers began to see that, as teachers, they can learn with and from each other.

Second, the preservice teachers decided *not* to respond to the whale-nami group in a way that would dismiss the students' ideas. Instead, based on whale-nami, they learned they needed to rethink how, in very concrete terms, they would place value on student thinking and support students' evolving ideas over time. To some extent, engaging in any teaching (micro-, macro-, or full lessons with placement students) involves anticipating how students will interact with the instruction, and developing preliminary instructional plans that flexibly and responsively bring together curricular goals with students' evolving understandings. In the context of macroteaching, the preservice teachers developed a preliminary unit layout (i.e., allocating a portion of the twelve hours of instructional time to eliciting students' ideas) but planned for flexibility in subsequent lessons to be responsive to the ideas raised. Even so, the ideas publicly proposed by the whale-nami group were unanticipated by the preservice teachers, and thus the preservice teachers were able to rehearse how to value the ideas present in the group's preliminary model while simultaneously "work[ing] on and with students' ideas" in their subsequent lessons.[12]

Third, from the teacher educator perspective, we were able to watch preservice teachers work through a complex pedagogical decision in real time. The whale-nami model disrupted the notion that all of the participants shared the same understanding of the science, and provided an opportunity for the preservice teachers to take actions to build relationships with their peers, rather than dismiss their ideas. Whale-nami became a named moment that the preservice teachers would reference in future lessons, including the fifth year of the certification program, as a primary example of how to recognize and value student thinking, thus building and sustaining relationships with students. The importance of the shared experience in a methods class cannot be understated: preservice teachers developed relationships with each other over the course of the two-semester methods course, collaboratively learned to notice and respond to student thinking, and began to develop a suite of core pedagogical practices as they learned with and from each other.

MACROTEACHING AS LEARNING OPPORTUNITIES FOR TEACHER EDUCATORS

As teacher educators, the goals we have for our preservice teachers in our practice-based methods courses extend past knowledge acquisition and the replication of discrete skills such as asking certain types of questions. Rather, we wish to support preservice teachers in developing the knowledge, practices, identities, and agency needed to enact AST in their own classrooms and to empower students to shift cultures of school and science. To support this type of methods course instruction, we need to identify evidence of where preservice teachers are on the pathway toward reaching these complex and multifaceted goals, and use that evidence to provide targeted feedback to preservice teachers. In other words, we need to find ways of using formative assessment (i.e., assessment *for* learning) if we want to support preservice teachers' abilities to enact AST upon entering the profession.[13] For us, macroteaching opened new avenues to think about formative assessment practices in our methods courses.

As teacher educators, we generally have the most insight into (and can provide feedback on) preservice teachers' planning and reflecting because those tend to be written assignments that preservice teachers submit for the course. However, the feedback we can provide on these assignments cannot target critical in-the-moment instructional decision making; rather, it occurs before or after a pedagogical teaching episode. During microteaching, we have short glimpses into preservice teachers' enactment of practices, and can sometimes provide immediate feedback. However, given our contextual constraints, we often have preservice teachers microteaching to their peers in multiple locations at any given time. Thus, as teacher educators, we cannot provide timely feedback to every preservice teacher. While examining video of preservice teachers' microteaching practice provides us with the ability to support post-instruction reflection, such analysis still does not allow us to offer instant feedback that is important for supporting learning of core practices.

Macroteaching, on the other hand, enabled us to target each preservice teacher's needs around learning to enact AST-based instruction as they were engaged in an authentic teaching experience. We were able to use our students' (i.e., preservice teachers') ideas, needs, and experiences to inform the formative feedback we gave them and the instructional decisions that we made around their developing practice. In some cases, we puzzled through decisions with preservice teachers as they worked through unexpected student ideas that arose in discussion.

For example, one macroteaching group in the third cohort taught a unit that used a documentary film about the drastic changes of the Rwenzori Mountains in Uganda as a phenomenon to introduce a unit on the water and carbon cycles, and human impact on Earth.[14] At one point in the unit, the macroteaching team was circulating to small groups and asking a series of preplanned back-pocket questions (such questions are aimed at pressing students to think more deeply about a phenomenon, and are often asked of each group).[15] One student raised the idea of gravity when responding to a question about the processes influencing the change in the mountains. One of the macroteaching team instructors initially dismissed this idea, saying not to "worry about that," and moved on to another student. However, Amelia Gotwals, the course instructor, called a "time-out" and asked the macroteaching group to pause and consider how they might take up this idea and further engage the student's thinking. During the time-out, members of the macroteaching team expressed uncertainty about what to do with the idea of gravity. After discussing possible next steps with the whole class, the macroteaching team called "time-in" and worked with the student to consider how gravity might be represented in his model. In this case, Amelia was able to pause instruction in the moment and provide feedback to the preservice teachers about using a student's idea, and the macroteaching team then moved forward in a way that was more responsive to the student's ideas.

Productive Floundering

During macroteaching, we, as teacher educators, sometimes gave explicit suggestions to the preservice teachers about how to consider a problem, such as in the gravity example. However, at other times we made the instructional decision to allow for "productive floundering," meaning that the instructional team faced and made difficult pedagogical decisions that sometimes resulted in confusion, frustration, or indecisiveness for the preservice teachers. When facilitated by teacher educators, such floundering can be productive, meaning that preservice teachers encounter a difficult situation, but—in the process of collaboratively working through the situation—develop a deeper understanding of AST and how to build and sustain relationships with students. In addition, preservice teachers learn how to build on their own instructional decisions, even if, in retrospect, those decisions may have led the class down a complicated path.

One example of productive floundering occurred with a chemistry macroteaching team during the second cohort. The chemistry team decided to use fireworks as a puzzling phenomenon to frame their macroteaching unit on chemical reactions. While this phenomenon had potential to frame the unit and elicit student ideas, the teaching group's launch of the instructional task (asking a vague question: "What do you notice about fireworks?") resulted in students' varied interpretations of the task and purpose. For example, some students tried to explain the colors during the explosions, some noticed and attempted to explain the size of the explosions, while others considered how the launch angle impacted the trajectory of the fireworks.

Given the students' varied interpretation of the task launch, the instructional team consulted with one another and the course instructors (David and Brian) to try to decide a productive pedagogical direction forward. While David and Brian suggested that the preservice teachers' attempt to bound the phenomenon by focusing students' ideas on the aspect most salient to their instructional goals, the instructional team decided that they wanted to keep almost all of the ideas on the table and find ways of addressing them in future lessons. However, as the unit unfolded, the group struggled to incorporate diverse ideas into a coherent sequence of lessons, and the students (i.e., their classmates) highlighted this feeling of incoherence in lesson debriefs.

Using feedback from their peers and teacher educators, the instructional team was eventually able to make iterative adjustments to their teaching in the concluding lessons to focus students' attention on key aspects of the fireworks phenomenon. This shift was made possible as the instructional team had time to reflect with their peers and teacher educators on the tensions they experienced between valuing all students' ideas and facilitating students' sensemaking of the fireworks phenomenon. The instructional team's reasoning process served as a useful window into the preservice teachers' learning, and allowed for substantive rounds of feedback that gave room for the instructional team to make pedagogical decisions that eventually led to meaningful shifts in their students' thinking and participation.

Reflections as evidence

In addition to providing an opportunity to gather evidence of and provide feedback on in-the-moment instruction, macroteaching also provided us, the teacher educators, with insights into preservice teachers' development of pedagogical decision-making. For example, during a lesson reflection, the group that taught the Rwenzori mountain unit addressed a comment from a student who said that they were unsure of how what they did in the lesson helped them to explain the puzzling phenomenon. One team member responded that while "this les-son may not make sense now, after the next two lessons, it should start to make sense." Her macroteaching teammates agreed and the class did not question the statement. However, the course instructor asked the whole class to reflect on a time when they did an activity in a science class, but were unsure of the purpose of the activity. The preservice teachers recalled how they often did not understand the purpose of science activities when they were K–12 students, and they felt that "doing science" simply meant completing the activity based on the criteria set by the teacher. The class then discussed how the macroteaching group could have publicly framed the purpose of the activity to better provide coherence for the students.

Overall, macroteaching allowed us, as teacher educators, to use preservice teachers' ideas, needs, and experiences to inform the formative feedback we pro-vided to them and to inform the instructional decisions that we made around their developing instruction. Having opportunities to provide feedback on pre-service teachers' instruction as they taught, and to use their daily experiences to guide our methods course teaching, allowed us to adapt our work as teacher educators based on evidence of their learning, thus representing what we hoped they would do with their students in classrooms.

CONCLUSION

As teacher educators who work to prepare teachers to enact AST in their own classrooms, we must find ways of providing preservice teachers with opportuni-ties to try out core practices, make pedagogical decisions in the face of inevi-table uncertainty, receive feedback about their progress, reflect on this feedback, and plan for future teaching. In the context of our five-year program, we were able to use macroteaching as an intermediary step in preservice teachers' learn-ing between microteaching and student teaching during a clinical experience in secondary science classrooms. Rather than oversimplifying the practices of AST, macroteaching allowed preservice teachers a safe environment in which to approximate complex professional work, such as working unanticipated student ideas into an entire unit of instruction.

Having a supportive group with whom to rehearse flexible and responsive instruction was critical for preservice teachers as they shifted toward elicit-ing, noticing, and using students' ideas in moment-to-moment and day-to-day interactions with students. Macroteaching provided us, as teacher educators, with a real-time and accessible window into the preservice teachers' reasoning and actions that we could use to develop specific and immediate feedback to support their learning.

By engaging in macroteaching as both teacher and student, preservice teachers had the opportunity to work through entire units of instruction based on puzzling phenomena and envision what AST could look and feel like in their future classrooms. Experiencing AST as students was important in highlighting the humanizing and relational aspects of the work. The preservice teachers learned the importance of building safe classroom cultures in which sensemaking was valued over simply stating a "correct" answer. Rather than experiencing AST as a set of practices that are sterile, disconnected, or prescriptive, macroteaching highlighted the importance of teachers and students codeveloping classroom communities in which equitable and intellectually rigorous work occurs throughout the school year.[16] Thus, we argue that macroteaching is one opportunity within practice-based teacher preparation that can provide preservice teachers with a strong foundation to design, enact, and adapt pedagogical practices that help students learn.

In our experiences, we have found that engaging groups of preservice teachers in macroteaching has provided an effective context in which to support their continued learning of AST beyond microteaching. However, we have lingering questions about how variations of enactments of macroteaching (based, in part, on our decisions as teacher educators) impact preservice teachers' learning. For example, macroteaching represents a cumulative learning experience for both the "teachers" and their "students." As such, we wonder how our choices of ordering macroteaching groups across the semester impacts the classroom community and learning experience:

- Which groups are best prepared to teach their unit early in the semester? Which group(s) might benefit from participating as "students" during one or more macroteaching units, prior to their own macroteaching?
- What do preservice teachers learn while participating in macroteaching as "students," and how does this support their teaching? Specifically, how will the decisions preservice teachers make in early enactments of macroteaching shape their peers' decisions during later cycles of macroteaching?

We also recognize that our two-year methods sequence is unique, and that many preservice teacher education programs include only one semester of science methods. Given these constraints, we wonder what features of macroteaching could be consolidated to fit within a single semester:

- Are there effective ways to elongate the microteaching structure in order to draw on particularly powerful elements of macroteaching, while still providing the time and scaffolds necessary for novice preservice teachers to learn AST?
- What affordances and constraints would such an expansion of microteaching (or alternatively, a consolidation of macroteaching) provide?

Last, we wonder about ways in which we, as teacher educators, can better support preservice teachers' framing of teaching as relational work:

- As teacher educators, how do we balance providing guidance on enacting core practices and allowing for potentially frustrating "productive floundering"? When does "productive floundering" cease to be productive and have the potential to negatively influence the relationships between teacher educators, preservice teachers acting as macroteaching instructors, and preservice teachers acting as students?
- In what ways does the relational work we do in methods classes, in particular during macroteaching, translate to the relational work preservice teachers do in secondary science classrooms?

Given these (and other) lingering questions, we are excited about the possibilities macroteaching affords, and are eager to learn more as the community of ambitious science teacher educators continues to develop and refine pedagogies of enactment to support preservice teachers' learning in practice-based preparation programs.

CHAPTER 6

Preparing Elementary Teachers to Support Science Talk

CAROLYN COLLEY AND MELISSA BRAATEN

The hallway outside Melissa's elementary science teaching methods class buzzes with nervous energy as twenty-six preservice teachers organize supplies and divide into groups for their first round of science teaching rehearsals. Melissa joins a group of five preservice teachers including Kate and Kevin. They are also joined by Sofia, a doctoral student with experience teaching multilingual kindergartners. Kate and Kevin start their rehearsal by telling the group of adults to play the part of kindergartners. Kate launches the lesson, introducing an anchoring context for learning about motion:

> Friends, look at this picture of Mr. Kevin going down the slide on our playground. Raise your hand if you have also gone down that slide. Good, it looks like everyone has been on this slide before. Turn to your buddy and tell them about what it feels like to climb up to the top and what makes your body slide down the slide.

After a minute of partner talk as kindergartners, Sofia calls a time-out, pausing the rehearsal to invite Kate and Kevin to deliberate about a dilemma of practice arising from this common routine of turn-and-talk.[1] Notice how this time-out sparks new ideas for both Kevin and Kate, allowing them to imagine a different way to orchestrate science talk with young children.

SOFIA: I want to think through how you got us talking and how that might play out in Kinder where everyone is still figuring out how to work together, to listen, to take turns, to be an equitable community. I also want to think about bilingual kids and how we can make talk happen for them, too. How could you use a tool to help us figure out how to participate together? How could you make sure bilingual kids can use all of their language resources to tell their stories about climbing and sliding?

KEVIN: Maybe we need like a partner talk card or something that helps kids remember that one person talks and the other listens and vice versa? Or, like, revisiting norms for sharing with and listening to a partner?

KATE: And I guess we could build in some wait time for thinking first. Maybe we could break it up like, "Imagine what it feels like to climb to the top—think of the story you want to tell," and, would I say something like, "Use whatever language you want," or is that not what I should do here?

After considering a few alternatives, Kate and Kevin decide that pairs should generate ways to listen to a partner and they will add think time for students to imagine climbing to the top of the slide and sliding down to help every student say more. Sofia invites Kate and Kevin to rewind and replay the opening of their rehearsal to try their newly imagined facilitation of students' science talk. After the rehearsal, the group engages in a reflective dialogue, adhering to agreed-upon norms to unpack student sensemaking.

This rehearsal structure—with time-outs, deliberation, rewinds, replays, and reflective dialogue with colleagues—provides learning opportunities for preservice elementary teachers about the interdependent elements of science lesson design, which hinge on teacher facilitation of elementary students' sensemaking through talk. In this rehearsal, Kate and Kevin recognize how their anchoring context leveraged kindergartners' everyday experiences of sliding down playground slides, giving all students access to the task. In addition, Sofia's time-out sparks an opportunity to design tools to support young learners' science talk. Finally, the reflective dialogue surfaces tensions inherent in engaging all students in purposeful science talk and helps preservice teachers address challenges together. Instead of dictating a "fix" for a problem (i.e., "You should do this"), the instructor guides the group to collectively inquire into a range of possible solutions to better support student sensemaking.

In this chapter, we describe how teacher education contexts can immerse preservice elementary teachers in the complex practices of facilitating learning opportunities that foster elementary students' science sensemaking. We share our course design and program features, which influence how we construct learning experiences for our preservice teachers. We also provide examples of activities we have found helpful for beginners to plan, enact, and reflect together through collective analyses of the *task*, *talk*, and *tools* rooted in Ambitious Science Teaching practices.[2] We conclude this chapter with next steps as we continue to learn how to prepare elementary teachers to support children's science sensemaking through talk.

OUR COURSE DESIGN

In a chapter focused on preparing elementary teachers to support science talk in school classrooms, why are we addressing our methods course design? From our experiences, we have found that we must design for and acknowledge the contextual and historical issues that surround and pervade our courses and the histories our preservice teachers bring into our classrooms. If we don't, preservice teachers pick up a repertoire of talk moves that invite students to talk, but they are not equipped to plan tasks, further facilitate robust classroom discourse, and design tools that support equitable and rigorous *sensemaking* talk beyond the initial invitation. Supporting productive talk with elementary students is not simply a matter of providing talk moves or questions, though these are a welcomed starting place for preservice teachers concerned with the immediacy of teaching in their field placements. The complexities of supporting classroom discourse for sensemaking is often overlooked when teachers discuss their planning processes, even though talking is one of the primary ways that people make sense of things together.[3] We design our courses to make visible the complexities of orchestrating sensemaking discourse and provide opportunities for preservice teachers to practice and reflect on students' science talk.

We teach at different institutions, yet our methods course goals are similar in that we focus on three main areas. First, we stretch preservice teachers' pedagogical imaginations to cultivate and refine a vision for what's possible for children's science learning experiences in elementary school—a vision that is often different from their prior experiences with science learning. Second, we support preservice teachers to recognize, name, and productively grapple with dilemmas together as a professional community. Third, to do this work, we introduce and use the Ambitious Science Teaching practices to support preservice teachers in planning for, facilitating, and reflecting on science learning experiences for children that are aligned with their newly imagined vision.

Table 6.1 summarizes the three phases of our courses and includes example assignments that we use to work toward our course goals. These phases are not a lock-step sequence; rather, they are cumulative and intertwined over our course timeline. We start by eliciting and working on preservice teachers' multiple and dynamic identities. As Thompson and colleagues describe in chapter 3, recognizing our own and others' worlds is central to developing critical consciousness about equity and justice in science education. Without acknowledging the histories and identities of our preservice teachers early in our courses, the learning opportunities, such as discussing examples of science teaching and engaging in teaching rehearsal cycles, are more tension-filled as novices seem less comfortable taking risks to try moves that deviate from how they learned science, what they think "learning science" means, and/or what they witness as examples of science learning in their school placements. We have found that our collaborative cycles of teaching rehearsals are more productive if preservice teachers have done some work around both eliciting their science identities (phase 1) and noticing, interpreting, and imagining work together (phase 2).

Phase 1: Eliciting and developing preservice teachers' science identities

We start here because our expectations for how elementary students participate through talk and the purposes for such talk in science classrooms are a dramatic departure from the norms and expectations for the kinds of science talk that we experienced when we were K–12 students; what it means to "learn" science has shifted over time. When we ask our preservice teachers to recount experiences they had as learners in science classrooms, one common theme resonates: science classes were often intimidating and even threatening. These quotes from two of our preservice elementary teachers exemplify this theme:

TABLE 6.1   Elementary science methods course goals, phases, and example tasks

ELEMENTARY SCIENCE METHODS COURSE DESIGN

Course Goals

1. Establish a collective vision for science teaching → Refine our shared vision
   and learning
2. Name dilemmas in elementary science teaching → Productively grapple with dilemmas
   and learning
3. Awareness of Ambitious Science Teaching (AST) → Design experiences using AST practices
   practices

Activity Phases

1. Eliciting Science Identities	2. Noticing, Interpreting, (Re)Imagining	3. Planning, Enacting, Reflecting
Where are our preservice teachers coming from? What matters to them and their identity as elementary science teachers?	What do preservice teachers notice in classroom examples? How do classrooms with productive talk support students? What dilemmas/questions do these examples raise?	What choices do preservice teachers make in designing and facilitating learning experiences? What dilemmas do these choices raise?

Example Tasks

Sharing Personal K–12 History	*Analyzing Classroom Examples*	*Teaching Rehearsal Cycles*
Share-a-Memory: What was a memorable K–12 science experience for you? Why did it stick with you? *Science & You Survey:*[a] Open-ended questions about personal relationship to and experiences with science learning.	*Watch/read examples of science talk* from elementary classrooms. Reflect and discuss: 1. What do you notice about talk? 2. How did students communicate ideas and sensemaking? 3. What makes you feel stuck/unsure as a science teacher? 4. What might you do differently as the teacher? Why?	*Reflective dialogue:* Use a conversation protocol with norms to foster reflective conversations at each stage of the rehearsal cycle. *Pause-rewind-replay:* Preservice teachers enact planned lessons in small groups, taking advantage of pausing, rewinding, and replaying to try out multiple pedagogical pathways.

Continued

TABLE 6.1 *continued*

Voicing Tensions and Dilemmas	Experiencing Lessons as a Student	Field-Placement Teaching
Sticky notes: What concerns do you have about teaching science? Sort by themes. Teacher educator uses themes to guide upcoming class session tasks. *Gallery walk/chalk talk:* Post questions about science teaching/learning around the room. Preservice teachers circulate and jot reactions.	Teacher educator teaches science with preservice teachers standing in for students for two purposes: 1. Embody an example of a teaching rehearsal cycle, including pausing to think aloud together about pedagogical pathways. 2. Allow preservice teachers to experience science learning through sensemaking talk about an anchoring event—likely different from their own school science experiences.	*Problematizing teaching:* Preservice teachers deliberate about lesson designs and pedagogical pathways before teaching, ideally with rehearsal before field-placement teaching. *Problematizing learning:* Preservice teachers analyze student sensemaking (seen in video and/or student work) and reflect on successes, tensions, stuck points, and next steps.

[a]Science & You Survey adapted from Okhee Lee and Cory Buxton, *Diversity and Equity in Science Education: Research, Policy, and Practice* (New York: Teachers College Press, 2010), 60.

- "I'm actually most nervous about teaching science because it's the subject I liked least as a child. I think my biggest challenge was getting bogged down and stressed about the 'rules' of science."
- "I know I am capable of being excited about science, but for so long science has felt like a chore to me, and that is not the attitude I want to bring to my future classroom."

Reflecting on their science learning experiences influences preservice teachers' visions of science teaching and learning by provoking images of the kind of teacher that preservice teachers want to be (or not) by replaying memories of what science classrooms looked and sounded like, and by remembering what kinds of student contributions were noticed, highlighted, and interpreted as worthy of attention by their past science teachers. We use tasks that elicit self-reflective and autobiographical experiences to bring preservice teachers' identities to the surface and offer new images of talk-based science learning experiences.

Preservice teachers also share their imagined futures as elementary teachers, signaling how much identity work is involved in becoming a teacher. These often highlight how difficult it is to envision ambitious and equitable science classrooms without having experienced them firsthand. The following quotes are from preservice teachers at the beginning of our courses, imagining a future that they had never seen or experienced:

- "To be honest, I am wondering what equitable science teaching even looks like."
- "I would like to cultivate a classroom community where equitable norms are established, and students know their power—that they have a voice and that it matters. Historically, like society, school has been a place designed for the success of some groups and the exclusion of others, and this is something I think we can combat. As an educator, I want my students to feel 'seen'; their experiences are validated."

Equitable science talk purposed at *sensemaking around ideas* is not typically part of our preservice teachers' personal K–12 science learning experiences, nor is it something that they typically observe in their placements at local schools as part of their teacher education program. This means that our science methods courses must provide experiences aimed at disrupting the patterns of discourse typically experienced in school science and shift expectations for science talk. Our preservice teachers make strides toward greater critical consciousness when they recognize and grapple with how they have been part of social and cultural systems that privilege some people and marginalize others. Seeing our own roles within systems paves the way for disrupting oppressive patterns of discourse and inviting new ways to talk science together. Such disruptions and invitations provoke deep identity work for preservice teachers who are reconciling their own past, present, and future identities as both science learners and science teachers. This phase lays the groundwork for phase 2, where preservice teachers see examples of science learning focused on students' sensemaking through talk.

Phase 2: Learning to notice, interpret, and (re)imagine elementary science

Learning experiences grounded in artifacts representing real-world science classrooms help shape preservice teachers' visions of purposeful science sensemaking talk. Artifacts like videos or transcripts allow for pausing, reviewing, and revisiting segments of student interactions, which is especially powerful for supporting preservice teachers as they learn to notice, interpret, and reimagine elementary science learning.

One task Carolyn uses for this purpose asks preservice teachers to reenact science talks using readers' theater–style transcripts, allowing novice teachers to experience, embody, and analyze examples of classroom science talk. In the next example, Carolyn adapted a published article that challenged assumptions about participation and meaning making when young children talk about the student-generated question, "Do plants grow every day?"[4] Notice how preservice teachers attend to multiple aspects of the elementary students' participation (phase 2) as well as reflect on differences in their personal experiences (phase 1):

Preservice teachers in Carolyn's elementary science methods course formed a fishbowl to reenact a second-grade science talk about how plants grow—half the class created an inner circle to perform the readers' theater script, becoming second graders, and the other half formed an outside circle, observing. As the inner circle began the reenactment, the mood of the room lifted. Adults erupted into laughter as they moved their bodies to mimic what they thought energetic elementary students might do to communicate their ideas about how plants grow. One preservice teacher even threw himself in the middle of the floor and extended his arms upward to act out plants growing with his whole body. Afterwards, preservice teachers discussed what they noticed about the students' science talk in small groups. Here is part of the conversation from Jenny's group:

JENNY: I work with kindergartners and I really like encouraging gestures and full body movement to help students share. I hadn't thought that students would connect to their growth chart or think plants grow more all at once.

BEN: Jenny's comment made me think about language support for students learning English, having to participate in English. Letting kids be kids and using drama.

LYUDMILA: I noticed that it felt light. Students enjoyed the talk and were free to contribute and excited—maybe *we* were acting more excited than the kids actually were [laughs]. But another thing I noticed . . . there was no evaluation or teacher correcting answers. Students said more and put out their ideas. That's different than the talk in school from when I was a kid.

BEN: I think students enjoyed the conversation because *they* asked the starting question and the teacher honored that and used it, so it came from them. But the teacher did not really do much. She only talked maybe two times in that whole science talk and let students talk about their own question. It seemed easy. Just let them talk.

ANDREW: One thing I noticed about this conversation was the lack of science content. I mean, students shared their ideas and they talked to each other, but I was thinking if I were the teacher, what could I do to give some science information? Otherwise, I don't feel like this conversation would really get anywhere. Did she [the teacher] move to another question the next week? [The article] said she does this kind of talk once a week—

JENNY: But I think that's okay—

ANDREW: I'm not saying it's bad, just . . . We can't—what if we did a science talk like this every day about a different student question? I'm not sure it would be chaos, but I don't know if students would actually *learn* anything.

JENNY: I think kids need this time so that students know how to talk to each other, and everyone feels comfortable adding their ideas and asking questions.

ANDREW: I don't know where I would go next from this science talk . . . so it's not just the kids with background or confidence talking. Like, how many other kids were in the room that we didn't hear from? And, like Ben said, I like that the teacher *wasn't in it* all the time, back-and-forth, and I like that the teacher wasn't evaluating anything, like what you said, Lyudmila. I'm still stuck on how to add content without taking over. Maybe if this is once a week, another day can feature a science concept so students can learn something new?

After small-group discussion, we debriefed as a whole class, allowing preservice teachers to notice multiple ways students communicate their ideas, consider the linguistic demands of talk, and raise dilemmas about how, if, or when to intervene with the power of the teacher's position and/or to inject a new content idea.

In addition to reenactments anchored in transcripts, another useful artifact to help preservice teachers envision elementary science learning opportunities is using classroom video. The bustle of live, real-time classroom observation makes pausing to reflect on parts of a science discussion challenging, if not impossible. Video clips of extended, unedited classroom interactions have affordances that make them particularly useful. Videos can be readily shared between teacher educators. Unlike real-time observations of teaching, viewers can slow down videos, revisit, rewind, and rewatch lesson segments to take up various interpretive lenses and play out the multitude of possible pathways created (or denied) in particular moments.

Viewing and discussing classroom artifacts together (phase 2) provides opportunities for preservice teachers to practice recognizing how their identities, beliefs, values, and professional visions inform what they notice, how they interpret moments, and what/how they contribute to discussions (phase 1). Preservice teachers are navigating and reconciling multiple, sometimes conflicting, visions of teaching and learning: their personal visions, visions shared by instructors in previous courses, and observations of mentor teachers at their school sites. These potentially contrasting professional visions present dilemmas to preservice teachers and teacher educators alike because our visions for teaching and learning inform how we view and analyze example artifacts. This is one benefit to revisiting and replaying segments of classroom examples using different interpretive lenses, focusing on opportunities for and evidence of students' sensemaking while giving space for preservice teachers to name conflicts, voice tensions, and raise issues to work through together in moments of discussion and in future opportunities for practice such as teaching rehearsal cycles.

Phase 3: Opportunities for beginners to plan, enact, analyze, and reflect together

Central to our goal of supporting preservice elementary teachers' facilitation of student sensemaking through science talk is a framework of three interrelated instructional design elements: science *tasks, talk,* and *tools.*[5] This task-talk-tools framework provides prospective teachers with scaffolding to help them plan, reorganize, or refine pedagogical tools and practices to keep children's contributions, ideas, and interests at the center of the complicated relational work of science teaching. Task-talk-tools framing offers preservice teachers a counterbalance to common elementary school sequences of fun but disconnected science activities devoid of intellectual depth.[6] The task-talk-tools framework scaffolds preservice teachers' consideration of the interdependent elements that support students' participation: Analyzing student *talk* creates conversations about *tools* that support student engagement in both *talk* and *task*, which often sparks revisiting and revising the intellectual demands for and purpose of *tasks*. This, in turn, spurs additional inquiries such as, "How do I choose the 'right' task for students?" Using cycles of teaching rehearsals with field placement enactments provides opportunities for preservice teachers to improve their abilities to recognize which parts of lessons can be salvaged and improved with revisions and which should be abandoned altogether, honing their senses of what blends of task-talk-tools best support students' sensemaking.

After analyzing classroom examples together over several class sessions, our preservice teachers expand their understanding of *task* beyond simply identifying *what* activity the students were engaged in to noticing *how* the teacher framed the intellectual demands of the task and analyzing how classroom communities work with and on students' ideas and science ideas together. Preservice teachers come to understand that examining student *talk* does not mean listening for vocabulary, correct answers, or certain talk moves, but rather, listening for how students compare, connect, refine, and revise ideas together. For some preservice teachers, repositioning the teacher from an evaluator to a facilitator creates a dissonance, which is why continuously reflecting on personal histories and values (phase 1) is so critical throughout our courses. Finally, preservice teachers expand their understanding of *tools*. Initially, many think we mean pieces of science equipment, such as a ruler or hand lens, but soon realize that tools include structures, scaffolds, and routines that foster engagement from all students in the task. To draw preservice teachers' attention to the power of sensemaking talk among children, we use a variety of tools to equip preservice teachers with a repertoire of moves that can be remixed and reorganized depending on different sensemaking situations with children.[7] Table 6.2 contains sets of questions we use to focus preservice teachers' analyses of science learning opportunities using the task-talk-tools framework.

The task-talk-tools framework is a useful analytical frame; however, it does not, on its own, provide a structure for preservice teachers to engage in cycles of planning, teaching, and reflection. To provide iterative learning opportunities, we anchor our coursework in a pedagogical routine called a *teaching rehearsal*.[8] During teaching rehearsals, preservice teachers design a short learning experience grounded in the use of science talk and then rehearse the orchestration of that talk with classmates and others as stand-ins for elementary students, as we saw with Kate and Kevin in our opening vignette about motion on a playground slide.

After each rehearsal, the group inquires together into dilemmas about teaching and learning that surfaced during the rehearsal. Reflective dialogues structured around dilemmas about teaching and learning are well regarded as productive learning opportunities for experienced teachers, but such deliberative experiences are not often incorporated into preservice teacher preparation.[9] For rehearsals in

TABLE 6.2   Task-talk-tool questions for designing, facilitating, analyzing, and reflecting on science learning experiences

Task	Talk	Tools
What were students asked to do?	What did you notice about student contributions?	How could tools (structures, scaffolds, routines) help all students engage in the ...
What was the purpose of engaging students in this task?	When was the talk focused on working on and with student ideas?	... intellectual demands of the task?
Whose ideas were featured?		
How were their ideas treated?	When did the talk seem most productive for students' sensemaking?	... communication and/or social demands of the talk?
Were all students able to participate in the task?		
Why or why not?		

Melissa's course, preservice teachers use a three-question protocol and adhere to the norm that participants do not "fix" problems by giving unwanted advice; instead, participants pose questions, listen, and paraphrase to learn together. These three questions are first posed to the teachers in the rehearsal (in our example, Kate and Kevin) and then to participants: (1) What did you notice about students' contributions?, (2) What are you wondering about science teaching and learning now?, and (3) What makes you feel stuck or unsure as a science teacher?

After rehearsing and reflecting with peers, preservice teachers modify and adjust their initial plans before moving forward to try out their science task, talk, and tools in their school-based field placement. In their school setting, Kate and Kevin then enacted their revised lesson with additional coaching from their mentor teacher to more productively facilitate students' sensemaking. This cycle culminates with a round of analysis and reflection using video and reflective dialogue about the classroom enactment of this science learning experience. Taken together, this cycle affords productive opportunities for preservice teacher learning. Preservice teachers need iterative opportunities to practice, analyze, and deliberate with colleagues about students' science talk followed by chances to decide on and try out adjustments in practice with children to develop into well-started beginner teachers.[10]

GRAPPLING WITH DILEMMAS

Despite our unwavering dedication to the foundational commitments of AST in our courses, two dilemmas continually surface in each of our teacher preparation programs. First, coursework focused on science teaching is located within a larger system of teacher preparation—a system within which courses seem to operate independently. These courses sometimes convey contradictory messages about teaching and learning, creating a challenge of curricular incoherence. Melissa's preservice teachers frequently note that in other courses they are guided to work on "content" and "language" as distinct learning goals, which provokes a dilemma when a focus on science talk works on content and language as intertwined learning goals. Carolyn's preservice teachers grapple with curricular incoherence in another way. Her program encourages using the same lesson plan template across all courses; however, parts of the template do not readily support planning for science learning aimed at being responsive to student sensemaking over multiple lessons. Preservice teachers grapple with how to use this template to "make it fit" for a multiday learning experience. These instances surfaced a lingering question for us as teacher educators: Should our framing and guidance to prospective teachers always align within our programs, or is there merit in articulating why different courses are offering different frameworks and guidance for preservice teachers?

A related dilemma of misalignment arises when the goals of teacher preparation coursework clash with preservice teachers' experiences in local schools, afterschool programs, and community-based organizations. This long-standing challenge is an example of the "two-worlds pitfall" where principles, policies, and practices may differ significantly between the "world" of teacher preparation and the "world" of field experiences.[11] Elementary schools serving as field sites for our preservice teachers have a high degree of variation in the time allocated to science each week, expectations for "what counts" as a science learning experience, and mentor teachers' orientations toward children's cultural resources for sensemaking. Furthermore, teacher preparation program schedules may or may not schedule field experiences to coincide with science methods coursework. That said, such dilemmas can be productive for teacher learning. We return to Melissa's context for an example:

> Toward the end of Melissa's course, preservice teachers completed their redesign of a set of science learning experiences from their school district's instructional materials. During a debrief conversation, one preservice teacher, Annie, and others in her cohort noted that they observe their mentor teachers using science activities downloaded from websites where teachers post, share, and sometimes

sell resources. They agreed that although children find these entertaining, they fall short of supporting students' sensemaking.

Later that week, Annie emailed Melissa to communicate that her mentor had, in fact, provided her with such an activity for her third, and final, teaching rehearsal and classroom enactment. Downloaded from Pinterest, a social media website used to share everything from recipes to craft ideas, this activity had students gluing different shapes of dried pasta to paper plates to show different stages in a butterfly's life cycle. Annie's email expressed her frustration: "There's nothing here to have a science talk about."

With help from her methods class community during her rehearsal, Annie developed an alternative plan, one that engaged students in making sense of animal growth and development focusing on the dogs, cats, and chickens that students' families raised at home. Annie presented this idea to her mentor, who agreed she could try it as long as she also completed the pasta craft. In her reflection, Annie communicated her newly developed stance as a science activity skeptic:

> During our carpet time science talk about pets and chickens, kids had a lot to say. Some kids clearly had seen puppies or kittens born, chicks hatching from eggs, and other kids had never seen those things, but they were so full of questions. There was so much there for them to talk about. But then I had to just basically cut it off like at fifteen minutes because we had to get the glue, the pasta, the plates, etc. and make this craft. The kids liked it, of course; they love making crafts. It was frustrating, but I learned a lot about being skeptical of whatever science activity people print and hand to me now. The science talk was much more worthwhile for students' learning.

Cases like Annie's illustrate how supporting elementary science teachers to create and sustain students' sensemaking talk in classrooms requires more than providing talk moves or sentence-starters or even showing what sensemaking talk looks like. Asking students to talk about a science concept devoid of context lacks opportunities for students to engage in deep explanation and sensemaking and limits participation. Therefore, in elementary science teacher preparation experiences, rehearsing science talk must be contextualized in order to provide authentic opportunities for teachers to learn and stretch their pedagogical imaginations. For Annie, the Ambitious Science Teaching framework served as a filter through which she critically examined the sensemaking potential of activity resources. Also, the teaching rehearsal cycle provided her with a collaborative space to refine her plan to support students' talk rooted in a genuine question about a real-world phenomenon relevant to her students. These experiences equipped Annie to reimagine and reorganize components of a science activity to strive for more robust sensemaking contexts for her students. Engaging in this cycle created a rich learning experience for Annie—one that she still talks about during every visit to her school. However, the contrast between Annie's professional vision for science teaching and that of her mentor teacher remains a troubling source of incoherence.

In Annie's case, there is a dissonance between the professional vision of the teacher education community at the university and the professional vision for science teaching at Annie's field site. In other cases, science teaching rehearsals and classroom enactments surface areas of incoherence within teacher education programs themselves, including between courses that focus entirely on lesson/unit planning without opportunities for enactment, analysis, or reflection and those that do. Equipping preservice teachers to grapple with dilemmas that arise when planning for and engaging in talk with students in a highly interactive science classroom requires careful interweaving of pedagogies of design, pedagogies of enactment, and pedagogies of analysis and interpretation. Incoherence between these teacher education pedagogies runs the risk of leaving preservice teachers frustrated and ill equipped for the complexity of supporting student sensemaking in the elementary science classroom.

OUR NEXT STEP: LEVERAGING LINGUISTIC RESOURCES

When asked to consider equity during discussions, preservice teachers readily recognize disparities between who speaks and who does not, varying lengths of contributions, and the powerful role of the teacher in determining which ideas and students are dismissed, ignored, or overlooked. Although these are all important considerations, we push our preservice teachers to expand beyond turns-of-talk as a marker of equitable participation, and to reconsider their role and power as teachers in disrupting or sustaining historical and institutional patterns of school discourse. Such reflexivity is essential for supporting preservice teachers to recognize their own and others' positionalities within science and school where power, status, and privilege structure inequalities. Deepening opportunities to develop critical consciousness is a high priority in our continued work with preservice elementary teachers. In recent iterations of our courses, we have focused more intentionally on how to elicit and leverage elementary students' linguistic resources for collective sensemaking.

To foster students' sensemaking talk, teachers must also foster a supportive social context within the classroom community. Prospective elementary teachers are acutely aware of the social risk involved with creating a talkative classroom where talk—especially tentative or exploratory talk—often positions students in higher or lower status, in more central or marginalized positions within the learning community. These social aspects of science talk have consequences that cannot be overstated. Without explicit tools, norms, and routines, students may not understand how to participate, feel limited in their ability to contribute, and/or have ideas but feel that their contributions are not valuable to others. For preservice elementary teachers to learn how to orchestrate science talks with children, they must also learn how to build and sustain generative social contexts.

To do this, we see a need for explicit opportunities to learn about and practice translanguaging pedagogies and other linguistically responsive practices during students' science discussions. Translanguaging means that multilingual students use their full linguistic repertoire to communicate ideas.[12] Designing from a translanguaging stance decenters English as the dominant language and the implicit (or, in some contexts, explicit) requirement of English proficiency to participate in school science. The following vignette from Carolyn's work in elementary schools highlights how central language and linguistically sustaining contexts are for students' science sensemaking:

> Second graders study photographs of apple trees and examine seeds before developing their initial models explaining how they think new apple trees appeared in a field that originally had one. Fifteen minutes later, students bring their models and gather at the carpet. The teacher invites Camila to start the discussion by projecting her model under a document camera for the class to see. Camila's writing is mostly in Spanish, with arrows and sketches showing a tree growing in size across several panels, each with an increasing number of branches, leaves, and roots, with raindrops and sun at each phase of growth. Camila points to what she wrote in Spanish, ad-libbing in English, to communicate her ideas about the importance of the rain and sun in helping the seed to grow bigger and bigger. She finishes, turns to the class, and asks: "Questions or comments?" Peter hops up and taps each arrow from left to right, asking, "What do these arrows mean?" She responds in English that each arrow means one year, and that this tree has ten arrows, so it is ten years old. She adds that the tree must get rain each year or else it will die. Other students nod. Peter sits down.
>
> Next, Nardos volunteers an idea in Spanish. Camila, Nardos, Luz, and a few others erupt into giggles. Micah asks for someone to repeat what Nardos said be-

cause he wants to know what was funny. Luz looks at Nardos, perhaps checking if he was going to speak. Then Luz faces the class and repeats Nardos's contribution in English, sharing that animals eat the apples on the ground, walk around, then poop out seeds when they have to go to the bathroom. Everyone laughs. This idea of animals pooping is now a simultaneously important and hilarious idea for these second graders. Vy holds up her paper to Nardos and points at her sketch of a cow showing the apple inside its stomach and a brown scribble on the ground. Her paper has several lines of Vietnamese writing and a few English labels including "cow" and "pop [poop]." Vy doesn't speak. Camila shows agreement by immediately grabbing a pencil and recreating Vy's drawing of a cow on her own paper, adding the apple inside the stomach and poop with seeds.

Capitalizing on the students' excitement about animal poop, the teacher prompts students to turn-and-talk about how they think this helps make new apple trees on the other side of the field. Students point at each other's models and use a mixture of languages and gestures. The discussion wraps with the teacher asking for help creating two lists, one capturing "Our ideas about what helps new apple trees grow" and another for students' questions, including: "What happens if *we* eat apple seeds?"

From this example, we could imagine how participation would be different if the teacher privileged English: Would the teacher have chosen Camila to start the discussion? Would Nardos have shared this important mechanism of seed dispersal? Would Vy's idea have been shared and taken up by others?

Some mentors and preservice teachers initially voice discomfort promoting multiple languages, citing reasons such as these: it makes English-only students and teachers uncomfortable, students might talk off-topic, and teachers have concerns about how to assess student learning in languages other than English. For these reasons and others, preservice teachers may initially feel pressure to actively make moves that maintain the English-only status quo of in-school science talk. These areas of discomfort and hesitation hint at underlying racial and linguistic ideologies that must be brought to light during teacher preparation if there is any hope for centering equity and justice in teacher education. As one starting place, we, as teacher educators, must seek and share examples of what this looks and sounds like in classrooms, which we have attempted here.

CHAPTER 7

A Possibility-Centric Vision of Elementary Teachers and Ambitious Science Teaching

CARLA ZEMBAL-SAUL, HEIDI CARLONE,
AND MICHELLE N. BROWN

The list of "troubles" associated with elementary teachers and teaching is long and likely familiar. For example: Elementary teachers' science backgrounds are limited; they are anxious about science teaching and often allow it to "drop off" the end of the day; when they do teach science, it is predominantly focused on memorizing vocabulary and doing "fun" but inconsequential hands-on activities. Deficit-based narratives perpetuate unproductive stereotypes about elementary teachers. As novelist Chimamanda Ngozi Adichie warned in her TED talk, "The problem with stereotypes is not that they are untrue, but that they are incomplete. They make one story become the only story."[1]

The era of Next Generation Science Standards (NGSS) elevates the importance of teaching and learning science in early grades, when children are naturally curious about how the world works and are capable of engaging in science practices. The contemporary vision for students' science learning and scientific practices demands radical shifts in instructional practices and renders all of us novice in some way, even those who previously excelled in science teaching and teacher education. As a result, most teachers will need professional learning opportunities aligned with the new vision for student learning. Professional learning will need to be coherently designed and intentionally build capacity among networks of teachers within and across grade levels, K–12. In light of this renewed cycle of learning for all, we consider elementary science teacher education across the career span—from teacher preparation, to induction years, to the continuous professional development of experienced teachers.

This moment of reform is an opportunity to revisit what we thought we knew, opening space for new vision, opportunities, and practices. We capitalize on this moment to encourage science teacher educators to reposition elementary teachers and the contexts in which they work as uniquely equipped, rather than woefully lacking, to enact science instruction that is rigorous, responsive, and just. What would elementary science teacher education look like if we flipped the script, from deficit dominant to asset oriented? It is this question that we explore in the chapter.

We label our vision "possibility-centric" because we want to draw attention to what *can be*. We begin the chapter by surfacing taken-for-granted assumptions about elementary teachers that describe the ways in which they are currently and have historically been perceived in terms of science teaching. These assumptions perpetuate unhelpful and sometimes damaging narratives of elementary teachers and constrain ways to organize their professional learning. Next, we map the goals of Ambitious Science Teaching (AST) to elementary teachers' strengths and the affordances of their school contexts. We move from foregrounding assets to proposing considerations for advancing a possibility-centric vision, and end the chapter with a call to action to engage in this brave work.

FLIPPING THE SCRIPT

To "flip the script" on deficit narratives, prevalent negative assumptions about elementary teachers and the teaching of science must be exposed, interrogated, and dismantled. Our ultimate purpose in this section of the chapter is to articulate the essential features of a vision that promotes assets elementary teachers bring and affordances of the contexts in which they work—a possibility-centric vision.

Confronting Assumptions About Elementary Teachers

Why does everyone say that elementary teachers don't know science? For decades, subject-matter knowledge and its transformation in support of meaningful learning has been central to research and practice focused on the nature, sources, and development of teachers' knowledge. Contemporary frameworks, such as AST, continue to recognize the importance of deep and flexible understandings of science, as well as science and engineering practices. Elementary teachers have taken the brunt of criticism about their science backgrounds, even though the literature is clear that secondary science teachers demonstrate gaps in their understanding that are just as problematic.[2] While limitations in science content knowledge are not really taken for granted when addressing elementary teachers, we need to stop admiring the problem and work toward innovative and powerful solutions.

Should we call elementary teachers "science teachers"? Using the phrase "elementary science teacher" is common in teacher education and the research literature. The issue is that most elementary teachers are not prepared as science specialists, nor do they identify as such. Applying a science-dominant label like "science teacher" inevitably positions elementary teachers at a disadvantage in the broader context of NGSS, and potentially AST. Moreover, the science education literature overwhelmingly disregards elementary teachers' multidimensional, contextual, and cultural histories. Even language as seemingly benign as "generalist" implies a wide but shallow array of knowledge and skills that overlooks the depth and intricacy of elementary teachers' work.

Positioning elementary teachers as having problematic science backgrounds and troubled identities dismisses their relational identities and the contexts in which they are inextricably bound.[3] By valuing only the science realm of elementary teachers' work, we diminish the complex understandings and abilities required to work with young children effectively and compassionately in elementary school contexts. Elementary teachers' agency in transforming their practice is inherently tied to their narrated and practice-based identity work and goals as teachers.[4] By recognizing and interrupting how researchers and teacher educators have narrated elementary teachers' identities through a science-norming perspective that is often misaligned with their historical and aspiring identities, we can widen our pedagogies to build on teachers' repertoires of practice. These perspectives challenge us as researchers and teacher educators to "see ingenuity instead of ineptness and inabilities, to see resilience instead of deficit, and to imagine futures . . . instead of imposing failure."[5]

What is the problem with silos in teacher education? A challenge unique to elementary teacher education is that teacher education programs are often designed around silos of early childhood, literacy, mathematics, social studies, science, educational psychology, and educational foundations. The assumption here is that teacher candidates will naturally identify important ideas and practices and make meaningful connections across discrete experiences. From an asset perspective, transcending silos is an opportunity to craft coherent curriculum to highlight commonalities, such as equitable access for every child to learn, caring and cultural formative assessment practices, and the role of productive discourse in sensemaking. An even loftier goal is to highlight important epistemic distinctiveness across subject areas, which is needed to address the larger aim of meaningful integration of the elementary curriculum. We point to Elizabeth Davis (see chapter 8) for a strong example of designing elementary teacher preparation in integrated ways that foreground high-leverage practices across content areas to minimize silos while maintaining epistemic distinctiveness.

Silos in teacher education also extend to pedagogical preparation and science coursework. In most programs, teacher candidates complete significant amounts of coursework outside colleges of education. Science coursework is typically taken in colleges of science with faculty who are not familiar with contemporary pedagogies aligned with AST for supporting teacher learning and development. While secondary science teachers take many more science courses than preservice elementary teachers, we know that more coursework does not equate to deeper understanding of science subject matter and science practices. Teaching science equitably and ambitiously means providing teachers at all grade levels and across career spans with meaningful opportunities to engage in science practices and sensemaking to support their continued development as professional educators.

If we consider teacher preparation as the responsibility of the entire university community versus solely education faculty, new possibilities for collaboration and innovation emerge. For example, at Pennsylvania State University, elementary education majors are required to complete three science courses as part of their general education experience—one each in life, earth and space, and physical science. Science educators have collaborated with colleagues in science and engineering to codesign and coteach a collection of courses tailored to meet the learning needs of preservice elementary teachers.[6] These courses intentionally integrate science discourse and practices with core ideas in science that K–5 teachers are likely to encounter in their professional lives. Future elementary teachers then enter teaching methods coursework and field experiences with a productive sense of the role of science practices, especially scientific discourse, in sensemaking. Science teacher educators can leverage this foundation as they pursue pedagogies to support the development of equitable and ambitious science teaching.

How inclusive is the meaning of "good teacher"? Most elementary teachers are white women and, because of the central role elementary schools play in cultural reproduction, white, heteronormative gender roles are the most valued and available for teachers in elementary settings. For instance, elementary teachers are often celebrated for their compliance, nurturing, and people pleasing, and also thrive from positive recognition from administrators who, too, are often hesitant to work against established norms.[7] There are also privileged ways to perform "teacher" that malign minoritized teachers.[8] Ladson-Billings provides a compelling case for why and how Black women are rarely held up as exemplary teachers.[9] For example, a "warm demander" style, often attributed to Black teachers who work with mostly minoritized, socioeconomically disadvantaged students, is a culturally responsive communication style that may look to be no-nonsense, severe, or overly structured to outsiders, but is a way to demand high expectations and communicate unwavering care for and belief in all students' abilities to succeed. Warm demanders' style of nurturing diverges from images borne of white, upper-middle-class values. It is not surprising, then, that warm demanders' ways to nurture are not part of historical constructions of a good teacher. Critically examining the hidden values of a "good" elementary teacher can surface inequities and lead to more just pedagogies for professional learning.

Normative definitions of "teacher" are simultaneously and historically rooted and produced in moment-to-moment interactions. In high-surveillance cultures, those whose teaching looks different from historical and local norms are increasingly at risk of being marginalized, ostracized, and even punished.[10] The literature about elementary teaching alludes to the hesitancy of teachers to take up practices that go against the grain and roles that violate gendered and raced models of "good teacher." For example, chapter author Heidi Carlone and colleagues spent two years in Ms. Carpenter's (a pseudonym) fourth-grade classroom, located in a school serving mostly minoritized youth. A Southern, white teacher from a self-proclaimed "poor, rural" background, Ms. Carpenter had a warm demander style that we sheepishly admit we sometimes found jarring. For example, she encouraged students to treat a class pet's death pragmatically because "that's what we would do on our farm growing up." She also resisted the normative teacher role by engaging her students with problems that administrators and other teachers deemed too advanced and not aligned with the curriculum; her curriculum was strongly student-driven. Her unconventional practices like regularly taking science outside and allotting time for all-day science explorations brought critique and derision from peers. Ms. Carpenter was professionally isolated, which was difficult.

In our work as elementary teacher educators, even those who deliberately teach against the grain express self-doubt about engaging in unsanctioned practices.[11] Historical meanings of "good elementary teacher" make equitable and ambitious science teaching difficult, but foregrounding them allows us to wrestle with realities left unspoken. Resisting narrow and racist definitions of "good teacher" could involve the creation of supportive, critically minded interinstitutional networks. For example, Kokka writes about STEM teacher activists who founded a social justice–oriented STEM organization because of their shared experiences of marginalization and quest for healing and empowerment.[12] Networks like these could facilitate teachers' navigation between racist historical and institutional realities to realize more expansive meanings of good teacher.

A POSSIBILITY-CENTRIC VISION

The previous section outlines implications of taken-for-granted assumptions associated with elementary teachers and their preparation and suggests ways to disrupt those assumptions. Here, we propose nonnegotiable features of a possibility-centric vision to advance equitable and ambitious science teaching practices based on elementary teachers' complex identities, assets, and resources, and the affordances of the school and community. Central to this vision is equity work, in which elementary teachers are uniquely positioned to engage. Equity work is ongoing, challenging, and intertwined with sociohistorical legacies of schooling and science that shape K–12 educators' work in every context, especially in vulnerable communities that do not have access to essential resources. Schooling's traditional practices can marginalize, oppress, and easily reproduce systems of oppression. Thus, equity work on the part of teachers and teacher educators is foundational to transformation aligned with rigorous, responsive, compassionate, and just science teaching.

In chapter 2, the authors assert that AST practices are intended to accomplish progressive goals for students' sensemaking that include bringing together elements of an emotionally supportive environment, productive discourse, engagement with the discipline, collaborative work, scaffolding, the appropriate level of intellectual challenge, and attention to equity—elements that some teacher preparation programs address in isolation from one another or discuss in abstract and disembodied terms.

While AST requires robust content knowledge, focusing on the aforementioned goals creates an opportunity to view elementary teachers as capable educators more broadly. Although there is more than one way to organize assets and affordances, what we have synthesized here is our take on the features of a possibility-centric vision for elementary teachers and the development of science teaching practices that are equitable and ambitious. We propose a strong orientation toward children's well-being; extended interactions in the school day and year to develop deep relationships with students, build trust, and allow for connections to their interests and lived experiences; potential to involve families in their children's education in meaningful ways; and the possibility for "generalists" to meaningfully integrate science across the curriculum.

Focus on children's well-being and development. Realizing AST in elementary grades cannot be achieved without attending to the relationship between teachers and students. Most elementary teachers pursue teaching because of their interests in nurturing children's social, intellectual, moral, and emotional development versus their interests in particular content areas. This is not to say they are not passionate about content; but often, their passion for children's welfare is central to their teaching vision. Elementary teachers' relational orientation toward their work, when examined through a possibility-centric lens, is a necessary aspect of science teaching that is equitable and ambitious. Their interest in nurturing the whole child may not lead directly to equitable science teaching, but it is a solid start upon which we can build. Windschitl, Thompson, and Braaten write: "[T]he most basic prerequisite for productive conversations is that all students *feel safe* in speaking out. A safe classroom is one in which students will not have their ideas ridiculed, and their teacher and peers will value what they have to say."[13]

Elementary teachers have extended time with the same students, allowing them to establish trusting relationships and create safe classroom environments. As mentioned previously, nurturing and caring are central to the identities of many elementary teachers and are essential to navigate students' socioemotional needs and development. Cultivating compassionate relationships with students is not only requisite for equitable and ambitious science teaching, but is also necessary for engaging students traditionally excluded from such participation.[14]

An excellent example of relational work that is both nurturing and rigorous is Mrs. Wallingdale's first-grade classroom in a school serving mostly minoritized students highly impacted by poverty. Carlone and colleagues spent two years in her classroom and were struck by the teacher's patience and skill in giving students time, space, and opportunities to create the meaning of scientific ideas with a wide range of literacy practices. Students' needs varied as they engaged in sensemaking. Some students needed to repeat others' ideas multiple times, some made bids to hold the floor for long periods of time, and others wanted to listen carefully in whole-group activity but later sought out adults to express their ideas during small-group work. The teacher knew her students, their needs, and their ways of thinking so well, she was able to see and leverage students' connections to science content that we struggled to see in the moment. Additionally, students felt comfortable helping others, frequently clarifying and/ or translating one another's contributions when there was confusion. By not shutting down their ways of sensemaking, Mrs. Wallingdale sent the message that all students' ideas, questions, and ways of expressing them mattered to the group's knowledge-generating activities.

Foster a sense of wonder through culturally relevant phenomena. A powerful affordance of the elementary context for advancing AST is that disciplinary core ideas and science practices are phenomena-based, informed by learning progressions research, and specified by grade level. Science lessons should no longer consist of one-off activities, but rather ongoing, question-driven, and coherently sequenced investigations of phenomena. Because children possess an inherent sense of wonder, compelling and relevant phenomena can incite interest, excitement, and persistence in "figuring it out." We caution, however, that not all phenomena are created equally. Teachers must seek to understand and draw upon students' funds of knowledge and situate content in questions and issues students care about, which is necessarily contextualized in family, community, language, and culture.[15]

In Zembal-Saul and Brown's work with predominantly white elementary teachers in a nascent immigrant community, teachers and researchers attend to the differences between conducting formative assessment of children's academic prior knowledge and seeking to understand their funds of knowledge.[16] We have come to understand that distinctions and intersections among prior knowledge and funds of knowledge are not intuitive to teachers, especially when working with culturally and linguistically unfamiliar student populations. Nevertheless, these understandings are vital to science teaching that is equitable and ambitious, and elementary teachers are clearly capable of taking up substantive practices for eliciting and connecting to funds of knowledge when supported in doing so.

Sensemaking in science is a collaborative effort. Student sensemaking and productive participation around complex science phenomena is a collaborative endeavor that is socially negotiated and dependent on shared knowledge. To do collaborative work, students take intellectual risks, challenging their own and others' thinking. The elementary context provides unique affordances for establishing collaborative norms and routines. Elementary teachers can explicitly model and have students practice working and interacting in groups, engage in listening to and respecting one another's contributions, and connect students' thinking to past experiences and learning to focus their attention on deep understanding of phenomena.

Elementary teachers' inclination to have students feel safe may lead to their resistance in challenging students' thinking and synthesizing knowledge. However, these teachers can overcome their hesitation and foster meaningful sensemaking with students through interactions between teachers and researchers.[17] In our work with elementary teachers, we have observed their abilities to create effective norms around collaboration and class discussions, and work with them to apply their skills to rigorous science sensemaking interactions. In some cases, these new norms disrupt traditional hierarchies so much that it is difficult to separate the roles of teachers and learners.

Research is clear that discourse is essential to sensemaking; however, the question of what counts as productive sensemaking discourse is not as tidy as the literature might have us believe. In practice, classroom discourse is wrought with meanderings, "off topic" contributions, power plays, and conflict. The tangles and uncertainties inevitable in classroom discussions may be quite foreign to middle and secondary science teachers who are accustomed to holding epistemic authority and controlling the flow of information through Initiation-Response-Evaluation (IRE) discourse patterns.[18] However, any primary grade teacher attempting to hold a relatively focused book talk knows quite well the challenges of facilitating productive classroom discourse and often has skills to pull out themes among the seeming cacophony of ideas. Although these same "generalists" may use IRE in other contexts, their skills with discourse can be leveraged and further developed for sensemaking in science. What if we started professional learning with elementary teachers assuming that they know a thing or two about how to manage the diversity of ideas in classroom discussions?

Opportunities to connect with home and family are accessible. Students often experience science instruction as separate from who they are, who they want to be, what they care about, their life experiences, and the resources and concerns of their communities.[19] Attention to equity in science instruction, however, demands getting to know our students, their families and communities, and the resources they bring to the classroom. While it is true that some families are, for good reason, disconnected from schooling as an institution, it is also likely that parents from all cultural backgrounds and socioeconomic situations are more apt to be involved with their children's elementary education rather than later years of schooling. Community-building activities, such as family science nights, are more common in elementary grades than middle or high school. Home visits, though still rare, also happen more frequently at the elementary level. A committed elementary teacher, then, can more readily prioritize connecting with families and communities, leverage those connections to expand what counts as knowledge in the classroom, and minimize gaps between school and home, as well as "science and me."

Building on the prior example from Zembal-Saul and Brown's longitudinal work in a community with rapidly changing demographics, engaging families and teachers in the education of emergent multilingual children is foundational. In addition to creating family science nights and school–home science investigations, teachers are beginning to seek out family and community members to inform their understanding of students' educational needs. In chapter author Michelle Brown's more than year-long collaboration with Ms. Matthews, a second-grade teacher, new immigrant family members increasingly expressed interest in participating in science learning opportunities. As families develop trusting relationships with Ms. Matthews, the possibilities for collaboration around culturally responsive teaching widen. For example, one mother shared how her daughter connected investigations of solids and liquids at school to cooking at home. By including families' voices, teachers gain opportunities to make deeper, more relevant, and equitable connections between home and school.

Integrate across the elementary curriculum. A potentially powerful affordance of the elementary school context for advancing AST is the potential to integrate across the curriculum. Most elementary schools are still structured around self-contained classrooms in which teachers are responsible for all subject matter. Language and literacy practices have been the cornerstone of elementary education for decades. Even we have lamented the lack of science in elementary schools because of overemphasis on literacy. When shifting from a problem-centric to a possibility-centric perspective, we can see this emphasis on literacy as a boon in the journey to enact science teaching that is equitable and ambitious. After all, the social practices of science and sensemaking require constructing, critiquing, and using scientific explanations and models through multiple literacies. Elementary teachers' pedagogical repertoires across domains of literacy (speaking, writing, listening, and reading) are assets to build upon in the pursuit of facilitating sensemaking.

It is clear from our work with elementary teachers and the research literature that they can learn to teach science in ways that are sensitive and responsive to the resources students bring to the learning community, leverage relationships with students and families, invite participation in productive sensemaking, increase student voice and agency in learning science, and more. In privileging elementary teachers' assets and the affordances of the elementary school setting, we can see possibilities amid barriers, resources over deficits, and can build on opportunities of context, rather than fixating on its problems.

ADVANCING A POSSIBILITY-CENTRIC LENS

In chapter 3, the authors make a strong case for engaging teachers in ongoing critical consciousness work to surface personal deficit perspectives and come to see all learners as having valuable resources for learning. As teacher educators and researchers, we too must uncover our deficit perspectives and shift our thinking and practices to build on elementary teachers' assets, as well as affordances of their school contexts. This is not to say that we should avoid areas previously perceived as challenging for elementary teachers; rather, we should reconsider the ways in which we support elementary teachers' learning and leverage assets to position them as capable professionals. Here we address four considerations grounded in our argument for adopting a possibility-centric vision for elementary teachers and their development of equitable and ambitious science teaching practices.

Treat Elementary Teachers as Capable Professionals

When elementary teachers' experiences are treated as legitimate resources, rather than handicaps, they begin to recognize and author themselves as belonging, and are able to engage in intellectual and pedagogical risk-taking related to science. They are positioned as cocontributors, as colearners, and inform and learn from others. In turn, their investment in their own and peers' learning increases.[20] The literature on elementary science teaching and approaches to professional learning designed for them have been slow to take up this stance. What would it look like to do so? We provide two suggestions.

First, we are intrigued by Bell's recommendation to shift away from focusing solely on "problems of practice," which inadvertently positions teachers' knowledge and practice as being in need of fixing. Instead Bell encourages identifying "problems and opportunities in practice," with the aim of supporting teacher educators to identify community assets that will help us do the creative, cumulative work of learning from one another and improving.[21] Shifts in language to consider the purposes of science teacher education for elementary teachers are more than just rhetorical moves; we know from learning scientists that language use is consequential in framing, organizing, and enacting learning.[22] The current political climate that increasingly deprofessionalizes teaching makes it difficult to move beyond challenges; a possibility-centric lens necessitates recognizing tensions between problem and opportunity spaces.

Second, we must provide space for elementary teachers to surface their multiple areas of expertise and explicitly connect that expertise to the skills, knowledge, and dispositions that are critical to teaching equitably and ambitiously. Elementary teachers' expertise is often rendered invisible, even to them. They view their long hours, thoughtful planning, efforts to develop relationships with families, relational work with children, and attention and care to the emotional, cognitive, and physical development of their students as nothing special. The contradictions in official narratives about what counts as expertise, exacerbated by high-surveillance cultures that diminish teachers' strengths and agency for improvisation and responsiveness, further dampen the recognition, by both teachers themselves and by others, of teachers as experts.

View Elementary Teachers as Collaborative Innovators—or Tempered Radicals

Tempered radicals "operate on a fault line," acting as innovators who succeed within the confines of institutional structures, but whose innovations sometimes position them as outsiders.[23] They are both leaders and valued collaborators who introduce new practices in their workplaces in incremental ways. Similar to Wenger's cultural brokers, they "manage carefully the coexistence of membership and non-membership, yielding enough distance to bring a different perspective, but also enough legitimacy to be listened to."[24] We need models of successful tempered radicals, an identity that demands bravery and resilience. Many of the tempered radicals we work with are teachers of color whose identities are not tied to normative, compliance, pleaser roles. Their ways of speaking up, advocating, and improvising are not always valued by others, but we can look for ways to use our voices and positions of privilege to amplify their voices and actions and support others in the tricky tempered radical identity work we describe here.

We highlight three lessons learned in our work with elementary teachers and preservice teachers to cultivate tempered radical identity work through science teaching. First, tempered radical identities are not final-form accomplishments; teachers' agency to author themselves as tempered radicals will ebb and flow throughout the school year. Attempts to study tempered radical identity work will be wrought with validity threats if one takes a "snapshot" approach. Practice-based professional development has to be carefully timed to align with teachers' readiness and agency to take professional risks. Initial forays into science teaching that is equitable and ambitious feel professionally risky. Building space to discuss and foster collective support for taking those risks should be a part of professional learning.

Second, tempered radical identity work emerges in practice. Low-risk opportunities to engage in coplanning, observing, and experimenting with AST practices are necessary for ongoing identity work as tempered radicals. In other words, teachers' identities and practices are iteratively connected. The practice-based approaches to teacher education described in this volume align well with cultivating tempered radical identity work.

Third, unyielding curricular schedules constrain teachers' creativity in exploring interdisciplinary connections and, in this arrangement, science time loses out. Teachers and administrators can seek ways to creatively include science as a regular part of the curriculum. This suggestion is a way to prod teachers out of perfunctory compliance modes. A tempered radical identity requires improvisation and imagination—"a process of expanding our self by transcending our time and space and creating new images of the world and ourselves."[25] That might mean quietly setting aside the day's reading time to investigate a critter that students find on the playground, or leading grade-level teammates to advocate to administrators for flexibility in time allotments when science instruction demands longer chunks of the day's schedule.

Build and Sustain Teacher Networks

Teachers who aim to enact tempered radical identities and teach science in equitable and ambitious ways run the risk of being marginalized for teaching "too ambitiously" and may be marked as troublemakers or disruptors. These kinds of shifts in practices thus demand more than bolstering teachers' knowledge of science and practices. Responsive pedagogies and professional learning demand allies and spaces for teachers' collective agency, where teachers cocreate tools for crafting bold new meanings of "good elementary teacher," support one another in enacting these meanings, and receive support from powerful stakeholders. Reform is never built on the shoulders of an individual, so we must create opportunities for teachers to strive collectively toward and be recognized and celebrated for their professional risk-taking. This is a big ask, but ignoring powerful historical legacies makes their reproduction more likely.

Teacher networks are one way to nudge elementary teachers out of comfortable pleaser identities, cocreate the tools and conditions to do so, and include administrators to bring about collective agency. The STEM Teacher Leader Collaborative (STEM TLC) is one example (www.uncgtlc.org). Carlone and colleagues founded this network with and for elementary teachers who work in high-needs schools to redress issues of STEM inequity and facilitate professional learning opportunities to learn to teach science and engineering equitably and ambitiously, reignite passions for teaching, celebrate teachers as professionals, and help them see and enact their teacher leader identities. Since teachers often work in isolation without structures and learning cultures that support collaboration, networks provide infrastructure to minimize professional isolation. Networks like these are popping up across the country, in physical and virtual spaces.

Bridge the School–University Divide

Beginning teacher socialization has proven to be a universal challenge for teacher educators to address. When even the most well prepared preservice and beginning teachers enter the workplace, there is a strong tendency to conform to that culture. This is the case whether the school culture and teacher practices are consistent with or radically at odds with preparation experiences.[26] It is not uncommon for experienced colleagues to explicitly convey to student teachers and new educators that they should "forget what they learned in college and focus on the *real world* of the school and classroom." This disconnect signals the need for a larger culture shift that focuses on disciplined inquiry into students' sense-making, continuous professional development, and collaboration. We urgently need to focus on building capacity among preservice and practicing teachers for teaching science equitably and ambitiously, as well as providing coherent professional learning experiences that intentionally feature the interconnectedness of theory and practice.

Professional development schools (PDS) represent a promising venue for creating professional learning communities with diverse membership focused on students and families.[27] At Pennsylvania State University, education faculty and teachers and administrators from the local school district have been involved in a long-standing, nationally recognized PDS at the elementary level.[28] With a coherent conceptual framework that utilizes the signature pedagogy of practitioner inquiry, and an emphasis on shared goals for student learning, the professional learning community engages in simultaneous renewal—all members have something to contribute and to learn through collaboration. Research from this field-based program demonstrates that teachers benefit from meaningful interactions with interns, university researchers, and graduate students engaged in learning about equitable and ambitious instructional practices, from observing and rehearsing these practices, and through the act of teaching and/or coteaching.

THE (R)EVOLUTIONARY WORK AHEAD

Our possibility-centric vision for elementary teachers and science teaching is hopeful. We acknowledge the complexities and constraints of problematic school policies and practices, as well as statewide mandates and testing, and we advocate for the need to support the continuous, coherent, and contextualized development of elementary teachers' science teaching that is equitable and ambitious. However, we do approach teacher learning from an essentially different standpoint: How do we provide support by surfacing and utilizing elementary teachers' epistemological, axiological, and contextual assets? It is altogether too easy to identify barriers; a possibility-centric vision asks us to flip the script in playful and imaginative ways. Doing so requires ongoing hard reflection about our own assumptions and approaches, coupled with patience with ourselves, to become aware of how we implicitly and intuitively position elementary teachers in deficit-based ways.

The danger of this argument is that we may come across as naïve to the very real barriers involved in rigorous, responsive, and just science teaching. To clarify, we are not here to paint a Pollyanna vision that blissfully and ignorantly wipes away gritty and sometimes demoralizing realities of elementary science teaching. However, in flipping the script toward the possibilities embedded in these realities, we begin to see and act differently. We act intentionally, with imagination and improvisation. Holland and colleagues' work is helpful; it acknowledges the overwhelming "webs of constraints that limit people's activities and life possibilities," but also the importance of paying attention to improvisations, "moments of resourcefulness," that allow for brave action, collective agency, and challenges to the status quo.[29] Our argument for lens-shifting, from problem-centric to possibility-centric, is a way to harness resources that already exist, if we can only learn to recognize and build on them. We imagine that this framing can be (r)evolutionary, pushing the boundaries of what's possible, and shifting as we learn with and from elementary teachers as accomplished educators.

CHAPTER 8

Practice-Based Elementary Science Teacher Education: Supporting Well-Started Beginners

ELIZABETH A. DAVIS AND JOHN-CARLOS MARINO

Elementary teaching is challenging! Elementary teachers typically teach all academic subjects, including mathematics, language arts, social studies, and science. Within science, elementary teachers teach all scientific disciplines—biological science, physics, chemistry, geology, and astronomy. Science is taught infrequently at the elementary level; schools tend to emphasize mathematics and language arts, and teachers may accept or even prefer that emphasis. Yet we know that even young children can engage in sophisticated scientific thinking and that they enjoy learning about the natural world—and that doing so positions them for success later in life. And we know, as shown in chapter 7, that elementary teachers bring many assets to the work of science teaching. Therefore, in this chapter, we focus on how we can support beginning elementary teachers in teaching science.

Over the last several decades, the fields of science education and teacher education have both shifted from valuing mainly conceptual knowledge and its application, toward the meaningful integration of knowledge and practice.

In science education in the United States, for example, we are moving toward "three-dimensional learning" that involves integrating disciplinary core ideas such as biodiversity, scientific and engineering practices, and crosscutting concepts such as size and scale, or "five-dimensional learning," which adds attention to interest and identity.[1] By *science practices*, we refer to the kinds of work in which scientists engage, such as asking questions, planning and carrying out investigations, and engaging in argument from evidence.[2] These science practices are used together and in combination with core disciplinary ideas and crosscutting concepts to support understanding natural phenomena.

In parallel, as described in chapter 1, teacher preparation has moved away from emphasizing (only) teachers' knowledge development and analytic skills toward *practice-based approaches to teacher education*.[3] These approaches support the development of both knowledge *and* practice. Knowledge is not separate from teaching practice, but an inherent part of the work of teaching. Here again, we see the importance of integrating knowledge and practice.

Table 8.1 summarizes three key meanings of *practice*, building on definitions within a study of teaching and a comparison to science.[4] How can teacher educators support beginning teachers in their development with regard to each of these meanings? As described in chapter 1, teacher educators can use pedagogies of practice, tools, and frameworks to support this learning.[5] Pedagogies of practice provide beginning teachers opportunities to learn to engage in the work of teaching. Tools and frameworks can support and extend novices' teaching by helping them to perform practices that might otherwise be out of reach.

In this chapter, we discuss how we use the pedagogies of practice along with tools and frameworks in elementary science teacher education, drawing on our own experiences as elementary science teacher educators, to work toward rigorous, consequential, and equitable science learning for every student.

WELL-STARTED BEGINNERS

The four-semester undergraduate program in elementary teacher education at the University of Michigan has always had a strong orientation toward content-area teaching and learning at the elementary level. The program includes purposefully designed clinical experiences for six to nine hours per week during the first three semesters of the program and a full-time student teaching experience in the final semester. The science methods class on which we focus here occurs

TABLE 8.1   Meanings of *practice*

	In teaching	*In science*
Used as a noun: A collection of practices	Core teaching practices used in planning (such as using curriculum materials) and interactions (such as meeting with a parent or leading a whole-class discussion)	Science practices used to learn about natural phenomena (such as building arguments or using scientific models)
Used as a verb: To practice, to rehearse, to work repeatedly on something	A preservice teacher may rehearse a lesson with peers before teaching it to children	A fifth grader may work repeatedly, throughout the year, on supporting claims with evidence
Used as a noun: A practice as in a profession	The profession is teaching	The profession is the discipline of science (e.g., biology, geochemistry)

Source: Based on Magdalene Lampert, "Learning Teaching in, from, and for Practice: What Do We Mean?," *Journal of Teacher Education* 61, no. 1–2 (2010): 21–34, doi:10.1177/0022487109347321; Anna Maria Arias and Elizabeth A. Davis, "Supporting Children to Construct Evidence-Based Claims in Science: Individual Learning Trajectories in a Practice-Based Program," *Teaching and Teacher Education* 66 (2017): 204–18.

in the third semester of the program. Typically, the class includes twenty-five to thirty preservice teachers. The first author, Betsy, is the lead faculty for the science methods class and teaches it most years. The second author, John-Carlos, has also taught the science methods class and served many times as a field instructor supporting preservice teachers in the program.

Betsy helped lead a significant redesign of the teacher education program in the early 2010s.[6] The program became more purposefully oriented around three pillars: a set of high-leverage, or core, teaching practices, content knowledge for teaching academic subjects in elementary school, and a set of ethical obligations for teaching. By content knowledge for teaching, we refer to the subject-matter knowledge and pedagogical content knowledge needed for teaching academic content.[7] Our goal is to support the development of "well-started beginners" who demonstrate beginning proficiency with a set of high-leverage, or core, practices, are "subject-matter serious" elementary teachers who are able to represent the content with integrity, and are ethical and equitable teachers who recognize and can act on their professional obligations.

How would these three pillars be reflected in elementary science teaching? To consider this, imagine a teacher teaching ten-year-olds a lesson exploring condensation—the change of state from water vapor to liquid water, with energy as an explanatory mechanism—and scientific modeling. Imagine that the teacher has students observe the phenomenon of condensation forming on the exterior of a can filled with ice water, and asks them to develop an initial diagrammatic model to explain the phenomenon. The teacher might need to be able to elicit students' ideas about the change of state from gas to liquid, adapt a lesson plan to meet the needs of their students, and use a routine for developing classroom norms for critiquing others' scientific models. These are some of the high-leverage, or core, science teaching practices in which they would need to engage.

The teacher would also need to understand the mechanism of the process of condensation (including its relationship to the crosscutting concept of energy), the typical ideas students may have about condensation or struggles they may have with scientific modeling, and the experiences with the phenomenon that could help them build on and from specific ideas. For example, the teacher would need to anticipate that students will think water has leaked through the can, and know that putting food coloring in the ice water will help students recognize the limited explanatory power of that idea. This is some of the content knowledge for teaching that the teacher would need.

Furthermore, the teacher would need to recognize ways in which students' models were and were not scientifically accurate, and move them toward more accurate representations (rather than lowering expectations and positioning some children as not being able to understand); engage *every* child in discussion and sensemaking (e.g., through using a mix of participation structures and careful scaffolding); and determine a range of examples from students' lives to make the science meaningful. These are some of the ethical obligations that would face the teacher in teaching this lesson.

PEDAGOGIES OF PRACTICE IN ELEMENTARY SCIENCE TEACHER EDUCATION

A study of professional education in three professions that rely on interactive, relational work—teachers, clergy, and therapists—found versions of pedagogies of practice in each.[8] *Pedagogies of practice* (or pedagogies of enactment) include decomposition (i.e., breaking teaching into its elements), representation (i.e., depicting teaching through videos or cases), and approximation (i.e., engaging in smaller or lower-stakes aspects of teaching). These pedagogies are powerful in supporting novices to do the relational and interactive work entailed in their professions, and to develop the knowledge and skills needed in scenarios like the one above using scientific modeling to explore condensation.

In the following sections, we describe how we have used each pedagogy in our elementary science methods course, along with associated tools and frameworks. We address "practice" with regard to both teaching practice and scientific practice, because of the central importance of both in learning to teach science, as well as how knowledge and issues of equity are intertwined with practice. To close each section, we pull in voices from some of our preservice teachers and graduates to give some insight into how we have seen these pedagogies support novice elementary teachers.[9]

Decompositions of practice

Decomposition of practice involves "breaking down practice into its constituent parts for the purposes of teaching and learning."[10] Learning the professions of teaching, the clergy, and counseling, for example, could include "focusing on the elements of lesson planning in teacher education, teaching aspects of speech and delivery for preachers, or targeting the development of the therapeutic alliance during the preparation of therapists."[11] Decomposition emphasizes the first meaning of *practice*, as noted in table 8.1 (i.e., a collection of practices, reflecting the ways work is done).

In our teacher education program at the University of Michigan (UM), we organize our overarching decomposition of teaching practice around nineteen high-leverage, or core, teaching practices. (See figure 8.1.) Our program developed this set of practices collaboratively.[12] In so doing, we used two sets of considerations. The first set is related to high-quality teaching. We wanted to identify teaching practices that were likely to be powerful in advancing students' learning, effective in using the differences among students to support learning and confronting inequities, and useful across many different contexts and content areas. The second set of considerations is related to high-quality teacher education. We wanted to identify teaching practices that could serve as building blocks for learning to teach, could be learned by a beginner, could be assessed, could be justified and made convincing to the preservice teachers and others (i.e., that had face validity), and were unlikely to be learned well only through teaching experience. The nineteen practices listed in figure 8.1 are our "best

FIGURE 8.1　High-leverage (or core) practices used in the elementary teacher education program at the University of Michigan

1. Explaining core content
2. Posing questions about content
3. Choosing and using representations, examples, and models of content
4. Leading whole-class discussions of content
5. Working with individual students to elicit, probe, and develop their thinking about content
6. Setting up and managing small-group work
7. Engaging students in rehearsing an organizational or managerial routine
8. Establishing norms and routines for classroom discourse and work that are central to the content
9. Recognizing and identifying common patterns of student thinking in a content domain
10. Composing, selecting, and adapting quizzes, tests, and other methods of assessing student learning of a chunk of instruction
11. Selecting and using specific methods to assess students' learning on an ongoing basis within and between lessons
12. Identifying and implementing an instructional strategy or intervention in response to common patterns of student thinking
13. Choosing, appraising, and modifying tasks, texts, and materials for a specific learning goal
14. Enacting a task to support a specific learning goal
15. Designing a sequence of lessons on a core topic
16. Enacting a sequence of lessons on a core topic
17. Conducting a meeting about a student with a parent or guardian
18. Writing correct, comprehensible, and professional messages to colleagues, parents, and others
19. Analyzing and improving specific elements of one's own teaching

bets" for a set of teaching practices that meet these considerations. We do not argue that these are the only teaching practices of importance, or even necessarily the best ones for a particular program to focus on; rather, we suggest that these are likely to be useful to consider.

Within our science methods class, we focus on a subset of these practices and work on them in the context of teaching science. These include:

1. Supporting students to construct scientific explanations and arguments (science version of UM practice #1)
2. Choosing and using representations, examples, and models of science content (UM practice #3)
3. Leading science sensemaking discussions (UM practice #4)
4. Eliciting and probing students' thinking about science (UM practice #5)
5. Setting up and managing small-group investigations (UM practice #6)
6. Developing norms for discourse and work that reflect the discipline of science, such as asking for evidence to support claims (UM practice #8)

Each of these has support in the field. Indeed, others have proposed core practices for secondary science that are close to these, despite the fact that we developed our practices initially in the context of elementary education and made them science-specific later.[13]

We also use another decomposition—an instructional framework to help preservice teachers decompose the work of *science* teaching, specifically. Similar to other models (e.g., the 5E model), it organizes the typical components of an investigation-based science lesson and names each component.[14] We call our version the EEE+A framework.[15] It decomposes science teaching into an *Engage* element in which the teacher establishes an investigation question and elicits students' initial ideas; an *Experience* element in which the teacher supports students in experiencing a scientific phenomenon and carrying out the investigation about it; and an *Explain+Argue with Evidence* element in which the teacher supports the students in making sense of the data, constructing claims based on evidence, and applying their knowledge to a new situation. (See table 8.2.)[16] We refer to the EEE+A framework as a *decomposition* because it plays a similar role as decompositions of core practices, by specifying key elements. It also serves as a *representation* of science teaching, by making those elements explicit.

The EEE+A framework links each of the main lesson elements to the teaching practices likely to be entailed in that element. Specifically, the *Engage* element typically entails the high-leverage, or core, practice of *eliciting student thinking*. The *Experience* element likely entails *managing small-group investigations* and choosing and using representations. The *Explain+Argue* element typically involves *constructing explanations and arguments*, choosing and using representations, and *establishing norms for discourse and work that reflect the discipline* of science. But the framework also connects to the science practices named in the Next Generation Science Standards (again connecting to the first meaning of *practice* from table 8.1). The *Engage* element entails the science practice of asking questions. The *Experience* element entails the science practice of planning and carrying out investigations. The *Explain+Argue* element centers on the science practices of analyzing and interpreting data, using mathematical thinking, constructing explanations, and engaging in argument from evidence. Thus, the EEE+A framework is intentional about the ways in which science teaching is decomposed, giving preservice teachers language for focusing on specific elements of the work, including the science teaching practices and the science practices.

TABLE 8.2   The EEE+A framework for investigation-based science lessons

Lesson element (overarching teaching practices in *italics*)	Likely dimensions of the lesson element (scientific practices in *italics*)	Relevant science teaching practices *Teachers may ...*
Engage with an investigation question (entails *eliciting student thinking*)	Establish an investigation question or problem (entails *asking questions*)	Pose or co-craft a question or problem for investigation. This question or problem should establish a meaningful purpose for experiencing the scientific phenomenon, and it should generate interest among students.
	Share initial ideas about the question or problem	Elicit students' initial explanations, models, or predictions to answer the problem or question. Encourage students to draw upon their prior knowledge and experiences.
Experience the scientific phenomenon to generate evidence to answer the investigation question (entails *managing small-group work, choosing and using representations and examples*)	Establish data collection for answering the investigation question or problem (entails *planning and carrying out investigations*)	Support students in setting up one or more investigations that allow them to gather data that they can use as evidence to answer the question or problem. With varying degrees of guidance, have students ... • Determine what data will be gathered and how and why it will be collected and recorded • Make justified predictions about the outcome of the investigation
	Carry out the investigation (entails *planning and carrying out investigations*)	Support students in systematically collecting and recording data (e.g., making scientific observations, making systematic measurements) to generate evidence to answer the investigation question or problem. This includes ... • Observing and listening to students as they interact • Asking questions to help students begin to make sense of what their data mean, rather than "telling" students the answer • Redirecting students' investigations to be more systematic, precise, and objective when necessary • Managing the distribution and collection of materials • Facilitating productive small-group work

Lesson element (overarching teaching practices in *italics*)	Likely dimensions of the lesson element (scientific practices in *italics*)	Relevant science teaching practices *Teachers may . . .*
Explain+Argue with evidence (entails *explaining core content, choosing and using representations and examples, establishing norms for classroom discourse*)	Identify patterns and trends in the data for answering the investigation question or problem (entails *analyzing and interpreting data, using mathematics thinking*)	Support students in making sense of the data so that they can generate claims with evidence. This includes . . . • Compiling class data and, if relevant, organizing or representing the data in meaningful ways (e.g., in tables or graphs) • Directing students to particular aspects of the data to help them identify and make meaning of patterns or trends in the data • Helping students select appropriate and sufficient data to use as evidence to support claims
	Generate scientific claims with evidence and reasoning (entails *constructing explanations, engaging in argument from evidence*)	Facilitate a discussion that enables students to answer the investigation question by using the data to generate evidence-based claims. Provide students with scaffolds, such as "I think ____(claim) because I observed _____ (*evidence*)" or "What I know: ____ (*claim*). How I know it: _____ (*evidence*)." You may wish to also support reasoning; for example, with "The science idea or principle that helps me explain this is _____ (*reasoning*). This helps me use my evidence to support my claim because _____." Provide opportunities for students to share their explanations with others, including peers, parents, etc. Help students . . . • Revisit their initial ideas about the investigation question, expanding upon or developing new evidence-based claims • Compare their own explanations with explanations reflecting scientific understanding, via direct instruction, textbooks, models, etc.; this includes introducing new terms to students, as appropriate • Question one another about their explanations
	Apply knowledge to new problems or questions	Support students in applying their knowledge to new learning tasks. For example, ask students "What would happen if . . ." so they can think through and explain their understanding of science concepts. Alternatively, give students a concrete new scenario that requires application of the new knowledge.

The EEE+A framework provides some support for novices in learning to scaffold students' full participation in science (see principle 3 in chapter 3).

As a decomposition of practice, the EEE+A framework is not just useful as a means for helping preservice teachers recognize high-quality science teaching or understand how the high-leverage, or core, practices might be enacted in the classroom. Preservice teachers can also use decompositions such as the framework to support their lesson planning and preparation when they enter the profession. One preservice teacher from our program, Claudia, continued to use the EEE+A framework after she left the program and began teaching on her own. In an interview, she said:

> I've actually used the sheet that you guys gave us that was kind of the cheat sheet of the EEE framework. And that's in the front of my science binder. I look at that to help me when I'm making investigation sheets. That's something the curriculum doesn't give me. (Claudia, interview first year, spring)

Indeed, when we viewed Claudia's lessons during her first year of teaching, the EEE+A framework was evident as the organizing structure, demonstrating that decompositions of practice can remain valuable even after teachers leave their preparation programs.[17] Claudia was similar in this way to other preservice teachers we studied, who credited the frameworks from the program, such as the EEE+A framework, in supporting their development.[18]

Representations of practice

The phrase *representations of practice* refers to "the different ways that practice is represented in professional education and what these various representations make visible to novices."[19] For example, teacher educators use brief narrative cases or accounts, video of classroom teaching, samples of student work, and lesson plans, among others. These different kinds of representations make different things visible, of course. A written case, for example, might help novices understand a teacher's reasoning, but might obscure what the teaching practice actually looks and sounds like. A video might show (some portions of) the classroom teaching practice, but would leave invisible the teacher's decision-making. Thus, representations can entail both the first and third definitions of *practice* from table 8.1 (i.e., a set of practices or a profession), though we mostly use representations for the first meaning.

Within our elementary science methods class, a key form of representation of practice is video. Video is crucially important in practice-based elementary science teacher education for three related reasons. First, elementary teachers, particularly new ones, often do not have much enthusiasm for teaching science and sometimes are nervous about doing so, as they frequently lack the rich subject-matter knowledge of their secondary colleagues.[20] Second, science is rarely taught at the elementary level, at least in the United States, taking the back seat to mathematics and language arts.[21] This means that there are relatively few opportunities for preservice elementary teachers to observe any kind of science teaching in their field placements or practica. Third, it is even more rare to see the kind of integration of science content and science practices recommended by current reforms. Activity-based science teaching is more common at the elementary level.[22] All three of these issues are exacerbated in urban contexts, leading to increased inequities in schooling experiences, particularly for students of color. As a result, we want to provide representations of high-quality science teaching practice to overcome these three issues and to work toward more rigorous and equitable experiences for children in a range of school settings.

To that end, we use videos fairly extensively. Typically, we look for videos that demonstrate the kind of rigorous, consequential, and equitable science teaching that integrates content and practice that we are urging our preservice teachers to engage in with their students. For example, the book *What's Your Evidence? Engaging K–5 Students in Constructing Explanations in Science* includes access to a set of videos showing the classrooms of Hershberger and other skillful elementary teachers engaging their students in such teaching, to illustrate the centrality of sensemaking.[23] We also make sure to sample across the grade levels our students will be teaching, including several at the lowest elementary grades, to highlight how young children are capable of engaging in sophisticated scientific thought and work.[24] Another characteristic that we look for in videos is the capacity to build a coherent storyline over a series of lessons. This can be important to overcome the one-shot nature of much of the science teaching that occurs in classrooms today; such a sequence can support novices in recognizing how lessons build coherently over time. Finally, because of the specific local context of the videos associated with the book, we also try to complement them with other videos that include students who more fully reflect the diversity of students in US classrooms. This is important because teachers may hold implicit biases about who is and is not a doer of science (discussed in chapter 3), though we see these perspectives change when challenged.

We learned over time that we need to use some kind of guide to support novices' engagement with video. Often, we organize this around the EEE+A framework discussed earlier in this chapter. For example, we use a standard observation guide, shown in part in table 8.3, to help orient preservice teachers to being able to recognize the elements of the EEE+A framework in science lessons they observe in the video representations of practice. (Another aspect of the observation guide focuses on issues of equity visible in the representation of practice.) Thus, representations of practice are often used in conjunction with decompositions of practice.

The observation guide reminds preservice teachers of the elements of focus (in table 8.3, the *Engage* element is shown), and then provides a space for preservice teachers to write notes about how the observed lesson (in the representation of practice) reflects that element of the framework.

Some of our work with preservice teachers in our program suggests the power of representations of practice in shaping their practice. For example, providing

TABLE 8.3   Observation guide based on EEE+A framework

Observation lens (overarching teaching practices in *italics*)	*Likely dimensions of the element or points* (scientific practices in *italics*)	*Relevant science teaching practices* **Teachers may . . .**	*How did this lesson depict the components of the Engage element?*
EEE+A Framework **Engage** with an investigation question (entails *eliciting student thinking*)	Establish an investigation question or problem (entails *asking questions*)	Pose or cocraft a question or problem for investigation. This question or problem should establish a meaningful purpose for experiencing the scientific phenomenon, and it should generate interest among students.	
	Share initial ideas about the question or problem	Elicit students' initial explanations, models, or predictions to answer the problem or question. Encourage students to draw upon their prior knowledge and experiences.	

video examples from younger grades can support preservice teachers in changing their expectations of young children.[25]

Approximations of practice

According to Grossman et al., "Approximations of practice refer to opportunities to engage in practices that are more or less proximal to the practices of a profession."[26] Examples include analyzing a written case or engaging in a role-play. Approximations span a continuum from less to more authentic experiences, often culminating in student teaching—still an approximation of full teaching practice, but reflecting many of its characteristics. Thus, approximations focus on the second meaning of *practice* shown in table 8.1 (i.e., to work to get better at something).

One key approximation of practice is rehearsal. Rehearsal involves "publicly and deliberately practicing" how to teach specific content using specific teaching practices.[27] Rehearsal can be used in conjunction with other pedagogies of practice to work in a purposeful way on developing specific knowledge and skill for teaching.[28] For example, in an analysis of rehearsals in which secondary science teachers were leading sensemaking discussions, we found that teacher educators could work with preservice teachers on important elements of teaching, such as attending to student thinking and to language.[29]

An example from our elementary science methods class is an assignment we call "peer teaching."[30] Peer teaching rehearsals entail having preservice teachers teach segments (organized around the three E's of the EEE+A framework) of a carefully selected lesson intended to highlight common problems of practice in teaching science, such as working with data gathered by children. They teach these lesson segments to a group of peers and a teacher educator. The teacher educator provides focused feedback. Preservice teachers have the opportunity to rehearse these practices in ways that "quiet the background noise" and lower the stakes.[31]

Because this elementary science methods class is positioned late in the program, the peer teaching experiences are designed to entail multiple science teaching practices (such as eliciting students' ideas about natural phenomena and supporting students in data collection), but are not full science lessons. Each peer teaching experience involves coplanning using an existing plan, enactment in small groups, and coreflection. For each enactment, a few preservice teachers work in parallel in the "teacher" role, working separately with a small group of colleagues as "students" as well as a teacher educator. In the enactment, the teacher educator facilitates in-the-moment feedback and, at times, recommends "stop-actions" or "time-outs" and "rewinds." The feedback is organized around the EEE+A framework and focuses on specific teaching practices. (See table 8.4; another section of the feedback tool supports a focus on issues of equity.) Feedback might provide the "teacher" some constructive criticism or might publicly highlight an effective instructional move. A stop-action might allow the preservice teacher to pause and regroup before continuing. In a rewind, the preservice teacher is given the opportunity to stop the action and try an instructional move again, in a different way.[32] For example, after stumbling over how to word a question about why hot water and cold water placed next to each other reach thermal equilibrium, a preservice teacher might be stopped by a teacher educator and asked to develop, and then ask, a better-formulated question.

In fact, the peer teaching experience (and other rehearsals) serve both as approximations of practice (for the individual who is teaching) and as representa-

TABLE 8.4   EEE+A *Experience* element peer teaching feedback focus guide

Likely dimensions of the lesson element	Relevant teaching practices	Focus questions to guide feedback— give specific examples
Establish an investigation question or problem	• Begins the lesson by posing or cocrafting a question or problem for investigation. • Establishes a meaningful purpose for upcoming activities and generates interest and curiosity among students.	How did the teacher support the students to establish an investigation question or problem?
Share initial ideas about the question or problem	• Elicits students' initial explanations, models, or predictions to answer the problem or question. • Encourages students to draw upon their prior knowledge and experiences. • Asks probing questions to encourage students to explain their reasoning.	How did the teacher elicit students' initial explanations to the problem or question based on their prior knowledge and experiences?

tions of practice (for the other participants, who are serving in the role of students). This dual role seems important in helping novices to notice student thinking.[33] Furthermore, the peer teaching also can help to decompose teaching, by zooming in on one or a small number of practices or subpractices. Thus, it reflects how the pedagogies of practice typically work synergistically with one another.

In our enactment of peer teaching, we use two different investigation-based science lessons from the local science curriculum. One, the Stems lesson, is aimed at the lower elementary grades and focuses on structure-function relationships in biology. The other, the Energy lesson, is aimed at the upper elementary grades and focuses on energy transfer and thermodynamics. This pairing allows us to work collectively with both qualitative and quantitative data, to look at change over short and long periods of time, and to work on different forms of data representation. Each lesson involves the opportunity to meaningfully integrate science content and practice, and each can be straightforwardly decomposed into the three E's of the EEE+A framework. (Our preservice teachers sometimes express that they wish we did rehearsals of lessons they are going to teach in the field, a feeling with which we are sympathetic; instead, we privilege using these two consistent lessons so our preservice teachers are able to see how different instructional decisions yield different opportunities to learn, and so the lessons can reflect the characteristics we particularly value, given our short time with them.)

In our course organization, we devote two weeks of class to each element of the EEE+A framework. The first is devoted to coplanning for the lesson element and the second to enacting the lesson element with one's small group. Then the process repeats for each subsequent element. See table 8.5 for a summary of how the structure works.

Because of the EEE+A framework's explicit attention to science practices, the peer teaching structure provides opportunities to support students as they engage in those science practices. For example, in the *Explain+Argue* element, preservice teachers work purposefully on two challenging aspects of science teaching: supporting students in analyzing and interpreting data and in constructing explanations and building arguments. By practicing this work in the low-stakes setting of peer teaching, they can work with colleagues on, for example, devising effective ways of representing scientific data to help students see patterns across it.

As with the decompositions of practice, preservice teachers can potentially bring their experiences in approximations of practice, such as the peer teaching

TABLE 8.5 Peer teaching structure

Timing in class	Activity for peer teaching	Who is involved	Expected outcome
Week 2	Get familiar with the lesson as a whole; coplan the *Engage* element of lesson	Each preservice teacher coplans with a group of other colleagues who are responsible for teaching the same lesson.	Each preservice teacher is responsible for developing a plan for the *Engage* element. The plans do not need to be the same across all members of the coplanning group.
Week 3	Enact the *Engage* element of lesson	Each preservice teacher is in a group of four, composed of two people responsible for the Stems lesson and two for the Energy lesson. They are all from different coplanning groups. One teacher educator joins each group.	Preservice teachers take turns enacting the lesson element for 15 to 20 minutes. There may be stop-actions and rewinds, as described in the text. After each enactment, the group reflects together, using a feedback tool organized around the EEE+A framework.
Week 4	Coplan the *Experience* element of lesson	Each preservice teacher coplans with the same group of colleagues as they did before.	Each preservice teacher is responsible for developing a plan for the *Experience* element.
Week 5	Enact the *Experience* element of lesson	Each preservice teacher is in the same "enactment" group of four as they were for the *Engage* element. Again, one teacher educator joins each group.	Preservice teachers take turns enacting the lesson element for 15 to 20 minutes, followed by structured reflection.
Week 6	Coplan the *Explain+Argue* element of lesson	Each preservice teacher coplans with the same group of colleagues as they did before.	Each preservice teacher is responsible for developing a plan for the *Explain+Argue* element.
Week 7	Enact the *Explain+Argue* element of lesson	Each preservice teacher is in the same "enactment" group of four as they were for the *Engage* and *Experience* elements. Again, one teacher educator joins each group.	Preservice teachers take turns enacting the lesson element for 15 to 20 minutes, followed by structured reflection.

activity, with them into their beginning years as a teacher. Jamie, a graduate of our program, reflected on a science lesson in an interview with us in the spring of her first teaching year: "You know, it's funny. I actually [often] think about . . . when we were teaching our lessons when we did our—when we split up into groups and we taught to the class. That's something that I have to go back to a lot."

She went on to describe how the peer teaching experience continued to help her consider ways to make her language clearer and more concise because "one of my tendencies is to be really verbose." The feedback that Jamie got during the approximation of practice not only helped her improve her teaching in the moment, but also pointed out an element of her practice that she continued to work on even after leaving the program.

Another graduate, Ginny, noted that it was the feedback she received during the peer teaching rehearsals that supported her development in eliciting children's thinking: "Well, with my peer teaching I didn't do very well at it. That was something I got a lot of feedback on— 'You should probably have some more questions to elicit student thinking and prior knowledge.'" (Ginny, interview, post-science methods)[34]

Ginny and Jamie were similar to preservice teachers who credited the approximations of practice throughout the teacher education program with supporting their development.[35]

In closing, we return to the several meanings of *practice* presented in table 8.1, and the exploration of those meanings with regard to both teaching practice and scientific practice.[36] In an elementary science methods course in initial teacher preparation, it is important to work on each of these.

The first meaning of practice refers to a collection of practices.[37] In this chapter, we have reviewed how our teacher education program at the University of Michigan is organized around a suite of high-leverage, or core, teaching practices, and how we focus on a subset of them in our elementary science methods class.[38] Many are interactional practices (that is, involving the interaction between teacher and students). This is in contrast to the more typical focus on planning practices or analysis and reflection in teacher education. By focusing on specific teaching practices, and not simply "teaching" in general, this sharpens the skills preservice teachers can develop.

The collection-of-practices meaning also includes the science practices.[39] We have shown how a range of science practices can be embedded into an instructional framework. By decomposing the work of science into these science practices, we can help preservice elementary teachers learn to support their students in engaging in those science practices.

The second meaning of practice refers to rehearsal.[40] In this chapter, we describe how we use "peer teaching" as an approximation of practice. This provides preservice teachers with scaffolded opportunities to rehearse the interactional work of teaching, supported by tools and frameworks. They can try out different wordings of questions, or different ways of organizing groups' collected data to support students' data interpretation. This is in contrast to more typical opportunities in teacher education to teach entire lessons, start to finish, without interruption. Though more authentic, a longer timeframe and lack of interruption eliminates the opportunity for the novice to gain focused feedback on specific teaching moves, a key feature of the peer teaching approximation of practice.[41]

Both children and prospective teachers need multiple opportunities to engage in science practices—to rehearse and become more proficient in them. By emphasizing the science practices throughout the peer teaching experiences, we illustrate how complex this can be. The preservice teachers come to see that "planning and carrying out an investigation" or "constructing a scientific explanation" is actually very challenging (for adult "students" and for children), and preservice teachers may come to appreciate the need for children's long-term, repeated engagement in this work, as well as how the science practices work together to support sensemaking.

The final meaning of practice refers to the profession.[42] By orienting the program around high-leverage, or core, teaching practices, content knowledge for teaching, and ethical obligations of teaching, we signal the serious work in which teachers engage.[43]

The EEE+A instructional framework we use in our science methods class, and the central focus on scientific practice more generally, help depict the authentic work of science.[44] Prospective teachers learn to take seriously their role not just in teaching science *content*, but also engaging children in science *practices* like modeling and argumentation. This, too, is a shift from traditional science teacher education, which has focused more centrally on preparing teachers to teach science content.

Elementary children should have the opportunity to integrate science ideas with science practices, but this happens infrequently in typical US elementary schools.[45] A range of opportunities to learn, including decompositions, representations, and approximations of teaching practice, can support beginning elementary teachers in learning to do this sophisticated work. By helping them develop a set of teaching and science practices, work repeatedly to learn to engage in those practices, and understand the professions of teaching and science—that is, by attending to the three meanings of *practice* with regard to both teaching and science—we can help preservice elementary teachers learn to support meaningful science learning for every student.[46]

CHAPTER 9

Preparing Teachers
in a Nonuniversity Site

ANNA MACPHERSON, ELAINE HOWES,
KAREN HAMMERNESS, PREETI GUPTA, NAINA ABOWD,
AND ROSAMOND KINZLER

The American Museum of Natural History (AMNH) conjures images of dinosaur bones, wildlife dioramas, and that iconic giant blue whale. People may not think of the museum as a context for teacher preparation and professional learning. However, cultural institutions have been playing an increasingly influential role in the educational landscape, and the AMNH is no exception.[1] The museum currently serves thousands of teachers a year in professional development programs and is also home to the only Master of Arts in Teaching (MAT) program occurring in a museum. The MAT program, situated in the Richard Gilder Graduate School, prepares approximately fifteen graduate students per year to teach in city schools. To date, the program has prepared 109 certified Earth science teachers to teach in high-needs middle schools and high schools in New York State.

Given the thousands of New York City teachers who visit the American Museum of Natural History each year, as well as its formal role in preparing new science teachers, the AMNH may be an ideal place to help develop and spread a common conception of good science teaching. But "What is good science teaching?" is not a straightforward question. At a place like the museum, where science learning experiences can take so many different forms (from free choice learning in the exhibition halls to formal classroom learning), the question becomes even more complex. Recently, multiple programs across the museum have explored the Ambitious Science Teaching (AST) framework as a way to help educators develop a shared conception of good teaching.

The AST framework, though not incongruent with the approaches used by many of the educational programs at the museum, does present a challenge to the dominant approach to teaching visitors about science in the exhibition halls. The AMNH is commonly seen as a place that provides "final form" scientific explanations, although it is working to change this conception. Objects and specimens are displayed in the halls accompanied by declarative text. Labels tend to answer questions, rather than pose them, though this is changing in new exhibitions. Volunteers who interact with visitors in the exhibition halls are called "explainers." It is challenging, then, to think about how to use the phenomena in the halls as things to be explained by visitors. While this is changing in newer exhibitions that focus upon opportunities for visitors to ask questions, make observations, or examine data, much of the formatting and labeling seem to contradict the work of sensemaking and the agency of the learner.

Museums are seen as sites of knowledge and as places of scientific and cultural authority.[2] Many visitors report that they visit museums to get their questions answered, to achieve a greater sense of understanding of the world. Yet visitors also report that museums and science centers can feel like a place "not for us"; and of course, museums do have a troubled history.[3] While many cultural institutions are making progress in addressing issues of equity, evidence also points to museums' history of systematically excluding diverse populations, and current research continues to reveal such inequities.[4] There is still far to go to ensure that museums are part of the solution, rather than perpetuating the problem. This discomfort with the authority of the museum and its troubled history mirrors the challenges raised in chapter 3 with science education and curricula that reflect dominant cultures and values, as well as language and interactional patterns. It is with this tension—between the museum as authority and the museum as a place in which people can engage in making sense of the world using their own knowledge as a resource—that we turn to AST as a framework for guiding our work with preservice teachers in the museum.

The focus of the chapter is two courses that help preservice teachers engage in the relational work of learning to interact with and support learners in sensemaking. These courses focus, in particular, on the practice of *eliciting and working with students' ideas*. We will identify the ways in which preservice teachers demonstrate increasing facility with enacting this practice, the main benefits we see, and the challenges. The first and second authors (Anna and Elaine) coteach EDU 620; the third and fourth authors (Preeti and Naina) teach in the Museum Residency course. "We" refers to either all authors or the instructors of the course being discussed, depending on the context.

THE MUSEUM AS SETTING

The MAT program was designed to address the shortage of Earth science teachers in high-needs schools in New York State. The preservice teachers in the MAT program all have a bachelor's degree in Earth science or a related field. Each cohort has approximately fifteen preservice teachers, composed of recent graduates and career changers. Of the 109 graduates of the program to date, 40 percent are people of color and 50 percent come from New York State.

The MAT's residency model places preservice teachers in high-needs partner schools where they coteach with classroom mentors. Preservice teachers complete a fifteen-month program during which they take a full load of graduate courses in pedagogy and complete two five-month residencies at partner schools. After graduation, MAT graduates are supported through two years of program-based induction.

Developing teaching practice across the program

Given the constraints of the MAT program, as well as the varied perspectives of the program faculty, preservice teachers learn about a variety of frameworks for science instruction; however, the AST framework, with its focus on student sensemaking, rigor, and equity, reflects central commitments and values of the MAT program for science teaching. Multiple courses (see table 9.1) offer preservice teachers opportunities to observe, analyze, and rehearse AST practices.

The three courses in the program that focus on AST are the summer museum-based teaching residency (Museum Residency), EDU 620: Curriculum and Instruction in Teaching Earth Science (fall), and EDU 675: Weather, Climate, and Climate Change (spring). This chapter describes the two courses that

TABLE 9.1   Courses in the MAT program that provide experiences with one of the four AST practices

	Residency	Course(s)	Focal practices			
Summer	Museum	Museum Residency	**Planning** around big science ideas	**Eliciting** ideas	**Supporting** changes in thinking	Drawing together **explanations**
Fall	School I	EDU 620: Curriculum and Instruction	**Planning** around big science ideas	**Eliciting** ideas	**Supporting** changes in thinking	Drawing together **explanations**
Spring	School II	EDU 675: Weather, Climate, and Climate Change	**Planning** around big science ideas	**Eliciting** ideas	**Supporting** changes in thinking	Drawing together **explanations**

▨ Course focuses on this practice.
☐ Course supports this practice.

focus on the practice of eliciting and working with students' ideas: the Museum Residency and EDU 620.

Museum Residency

The Museum Residency provides preservice teachers with an opportunity to observe and teach in a "low-stakes" setting, meaning both the learners and the teachers cannot "fail" in the experience. The objectives of the course are to contribute to preservice teachers' understanding of how people learn and how that learning is mediated by multiple factors. including culture, gender, and age; to help them become familiar with how informal science institution resources (for example, museum exhibits and objects) can be used for motivation, engagement, and demonstration of key ideas; and finally, to help preservice teachers develop teaching practices, such as eliciting and working with students' and visitors' ideas.

Throughout the summer, preservice teachers are divided into teams and rotate through experiences in museum programs. These experiences include facilitating discussion at educational touch carts, facilitating in a children's discovery room, observing during a two-week middle school institute, and designing learning experiences for a one-week high school institute. This scaffolded set of experiences positions them to gradually take on more teaching responsibility over time, from holding one-on-one conversations with museum visitors, to conducting in-depth classroom observations of students in AMNH's out-of-school programs, to codeveloping and coteaching a short Earth science lesson at the end of the summer.

In particular, working at the touch carts in the exhibition halls allows preservice teachers to develop their practice of eliciting and working with visitors' ideas. The carts have objects relevant to the content of the hall, available for visitors to touch (unlike most items in the museum) and talk about. Preservice teachers spend twenty hours working at the carts in five different halls, including, for instance, the Hall of the Universe and New York State Environments. With each interaction, they practice how to engage the learner, access prior knowledge, introduce the big idea, and assess for understanding. In this context, they work with people of varying levels of English language proficiency and prior science knowledge, even within a single family unit.

To prepare for their work on the carts, preservice teachers observe experienced cart facilitators and spend time learning about the objects. They also participate in activities and research discussions about how people learn, and the role that prior knowledge and talking play in learning. For example, course faculty introduce candidates to conceptions of prior knowledge through an activity involving melting ice cubes in salty and fresh water. Preservice teachers make a prediction about which ice cube will melt faster. Then, they observe what actually happens (which often contradicts their initial prediction), and have a conversation about how prior knowledge influenced their experience during the activity and their understanding of what actually happened. Together the class reads a summary of research about the idea of prior knowledge and discusses it in relationship to the activity and the theoretical principles.

In addition to exploring prior knowledge, preservice teachers rehearse "intergenerational conversations." As explained in chapter 1, "rehearsal" is a teacher education pedagogy that approximates the work of teaching by providing a space for novice teachers to open up their instructional decisions to one another and their instructor.[5] In preparation for rehearsal, preservice teachers learn strategies for facilitating a conversation with a group of people of different ages. They focus on how to engage members of these families in talking to each other, rather than solely responding to the facilitator. They also rehearse questioning strategies.

Then, for the actual rehearsal, the novice teacher practices with two visitors three separate times, first practicing lecture; then IRE (Initiation-Response-Evaluation); and finally, trying out the use of open-ended questions. They then analyze each of the three rehearsals, debating the benefits and drawbacks of each question type.

Preservice teachers videotape themselves at least once while on the carts. They are divided into small groups of "critical colleagues" who offer feedback to each other on their cart facilitation. For many, this is their first opportunity to watch themselves teach and receive feedback. To that end, we put a lot of effort into making the experience safe and comfortable. The presenter identifies a specific "area of focus" around which they want feedback. This reinforces the idea that the presenter is in control of the experience. Critical colleagues then discuss what they saw in the video and offer feedback to the presenter using a protocol. Feedback is focused on the teaching practices of eliciting visitor thinking, questioning, and drawing out the "big ideas" of the cart. Though preservice teachers go into this experience with a lot of discomfort and sometimes anxiety, they almost universally emerge from it with positivity, and describe it as a valuable opportunity to reflect on their own practice.

What preservice teachers learn during their carts rotation. During their experiences on the carts, preservice teachers begin to realize that although they may be experts on a given topic, they do not need to demonstrate all of their expertise to visitors. Rather, they can leverage their expertise to plan an experience that will engage visitors and set them up to seek additional information. While they begin relying almost exclusively on facilitator-centered discourse, over the course of the carts rotation they start to shift toward visitor-centered discourse.

Furthermore, preservice teachers discover that open-ended questions can shift the discussion to be more learner-centered. For example, they begin to notice that closed-ended, recall-only questions (e.g., "What is this?" or "What do we call this?") are often met with blank stares and lack of engagement. MAT faculty remind them of what they have learned about questioning strategies and how it is easier to engage visitors with questions that are based on observations visitors can make, such as "What do you notice about these two rocks?" While preservice teachers do see the difference in the interaction when they start with more open-ended sensemaking-based questions, they struggle to maintain that practice. This tendency to default to recall-only questions is reinforced when visitors to the museum believe they are supposed to "listen" to the presenter only, rather than share their experiences and questions. Uncomfortable with the silence, the preservice teachers get nervous and revert to the "what is" question. The cart experience is where preservice teachers begin to notice this challenge of shifting discourse to be more learner-centered. Courses later in the program are designed to provide additional opportunities for preservice teachers to develop their practice in this regard. For example, in their curriculum and instruction course (see the next section), they are introduced to a set of "discussion moves" to develop their ability to sustain a learner-centered discussion.

Finally, their experiences on the carts show preservice teachers that all visitors come with prior knowledge and rich experiences. For example, many are shocked that young children know so much about certain topics. Many preservice teachers at this stage in the program report a new commitment to not making assumptions about learners, a finding that is consistent with existing understanding of revelations that floor facilitators in other museums have about engaging visitors.[6]

EDU 620: Curriculum and Instruction for Teaching Earth Science in Secondary Schools

EDU 620 is designed to prepare preservice teachers to teach Earth science in middle and secondary schools, and emphasizes curriculum and instruction that support students in learning Earth science core ideas and science practices. Additionally, preservice teachers learn about core science teaching practices with the goal of meeting the needs of all students in their science classrooms. This course focuses on three primary science teaching practices: planning for engagement with big ideas in science, eliciting students' ideas, and pressing for evidence-based explanations.

Building on their experiences in the Museum Residency, EDU 620 particularly focuses on supporting teachers in eliciting students' ideas, a practice that novice teachers often struggle with. New teachers often come into teaching focusing on what they do as instructional designers, or what "they do" to teach. As we observe every summer in the Museum Residency, the shift to pay attention to the learner is difficult for novices to make. They have learned that science class is a place to show what you *know*—and the right things to know originate with the cart facilitator or teacher. However, we aim to help preservice teachers recognize that it is impossible to plan meaningful instruction without surfacing students' ideas about the phenomenon; thus, we focus on "eliciting" early and often.

One way in which we work on the eliciting practice is through an assignment called "Eliciting students' ideas with observable phenomena." In the cycle for collectively learning to engage in ambitious instructional practice, this work would be considered a type of sheltered planning and practice—preservice teachers collaboratively plan with peers and conduct a rehearsal.[7] Small groups of preservice teachers choose an Earth science–related demonstration that illustrates a key phenomenon from a list recommended by course faculty. For example, one phenomenon on the list is the "can crush," in which a soda can is heated on a hot plate and then is violently "crushed" by placing it into a bath of icy water, illustrating the relationship between the temperature and pressure of a gas. This type of observable phenomenon could function as an "anchoring phenomenon" around which an entire unit could be planned. It could also be used as a "lesson-level" phenomenon to help students figure out a smaller piece of the larger picture. We distinguish between these two possibilities; however, in both cases, an "eliciting" phase is required.

Candidates are provided with a set of strategies to support their planning—they are encouraged to think about their goals for the "eliciting" phase, what their students may know already, and what key scientific concepts are related to the phenomenon. They are provided a list of "discussion moves" they can use to sustain the discussion, and are encouraged to think about where they will incorporate wait time.[8]

Before candidates themselves take responsibility for teaching their classmates, the course professors model the strategy twice to provide preservice teachers with images of what this core practice could look like in a science classroom. During these initial sessions, preservice teachers are introduced to the practice of analyzing teaching. In an effort to build norms—an important aspect of a productive rehearsal—around talking about peers' instruction in an analytical but nonjudgmental way, we focus the "debrief" around the use of discussion moves. We organized the "moves" into a chart and copied it on brightly colored paper as a way of creating a signal—that is, when we are talking about each other's teaching, we have our "discussion moves" chart out. We deliberately frame our comments in terms of how discussion moves were leveraged and what the result of those choices was, rather than, "Oh hey, that was great." A section of the "discussion moves" chart is shown in table 9.2.

After recognizing over the course of teaching that preservice teachers often struggled to pace the twenty minutes of instruction, we further scaffolded the assignment by providing suggested thinking routines to structure students' observations of the phenomena, connections to their lives and similar phenomena, and initial science ideas that might help them explain the phenomena. For example, preservice teachers could use the Prediction, Observation, Explanation routine, with the final phase renamed "preliminary explanations."[9] In this routine, the teacher prompts students to make a prediction about what they think might happen. During and immediately following the demonstration of the

TABLE 9.2   Discussion moves in science

Discussion move	Definition	Example in science	Evidence of use
Marking	Responding to student comments in a way that draws attention to certain ideas.	"I want to focus on something Charlene said because it's important. She said that the temperature of the object makes a difference in whether condensation will appear."	
Turning back to students	Turning responsibility back to the students for figuring out ideas.	"How does that connect to what we already know about ecosystems?"	
Turning back to text	Turning students' attention back to the "text" to clarify or focus their thinking. In science, the "text" may be a data table, a graph, or a model, in addition to what we usually think of as a "text."	"Does the data table tell us about when the growth of the roots and the shoots was happening the fastest?"	
Revoicing	Interpreting what students are trying to express and rephrasing the ideas so that they can become part of the discussion.	"So you're pointing out that the wire, the battery, and the bulb have to be connected in a very particular way in order for the bulb to light."	

Source: Based on S. Kademian and E. A. Davis, "Planning and Enacting Investigation-Based Science Discussions: Designing Tools to Support Teacher Knowledge for Science Teaching," in *Sensemaking in Elementary Science: Supporting Teacher Learning*, ed. E. A. Davis, C. Zembal-Saul, and S. Kademian (New York: Routledge, 2020), 113–28.

phenomenon, students are prompted to make observations. Then, the teacher facilitates discussion about what the observations mean and students collectively construct an explanation for the phenomenon. Though used by preservice teachers less frequently, we also suggested A/B Partner Talk[10] and See, Think, Wonder[11] as possible thinking routines. Selecting a thinking routine helped preservice teachers think about their goals for student talk, and the routines succeeded in helping them pace instruction.

ANALYSIS OF REHEARSALS OF "ELICITING STUDENTS' IDEAS"

Two years of developing the focus on "eliciting students' ideas" has led to encouraging results: preservice teachers frequently report that they found this work valuable, clinical faculty report seeing the practice enacted in residency classrooms, and graduates report experimenting with the "eliciting" approach. For example, during observations of graduates participating in the induction program, faculty reported seeing teachers begin units of study by eliciting ideas from students about an observable phenomenon using the Prediction, Observation, Explanation and See, Think, Wonder routines. We became interested in how the rehearsals were functioning to develop the preservice teachers' practice so that they were doing the kind of work we were hearing about. Thus, in 2018–2019 we chose to videotape the rehearsals to help us better understand which discussion moves preservice teachers were using and in what ways/circumstances. Were some approaches to eliciting students' ideas more easily taken up? Were others more challenging and less likely to appear in their teaching? We particularly wondered how preservice teachers developed their thinking about and comfort with leaving final explanations of phenomena for later. We also wondered how the rehearsals might be helping them develop an analytical stance toward teaching practice.

The rehearsals conducted in each session are summarized in table 9.3. In each rehearsal, preservice teachers conducted twenty minutes of instruction during which they (1) demonstrated the phenomenon, (2) elicited students' ideas about what they saw, and (3) prompted students for everyday experiences that connected to and might help explain the phenomenon. Following the rehearsal, there was a structured debrief in which the teachers reflected on their planning, the designated observers offered observations connected to the "discussion moves" chart, and the instructors offered feedback.

TABLE 9.3   Summary of the "eliciting ideas about observable phenomena" segments in each session of EDU 620

Session	Teacher(s)	Phenomenon	Scientific concept; Earth science connection
1	Anna	A wooden meter stick is smashed in two by a single sheet of paper.	Air pressure; weather
2	Elaine	A can of Diet Coke floats; regular Coke sinks.	Density; composition of earth; weather
3	Stephanie, Grace, Courteney	Colder water (dyed blue) sinks; hotter water (dyed red) rises; the water swirls.	Convection; ocean currents; weather
4	Hannah, Keya, Morgan	Blowing bubbles into water with BTB turns the water yellow.	Gas dissolving in water; ocean acidification
5	Kate, Linda, Bree, Sara	A long plastic bag fills completely with a single puff of air.	Bernoulli effect; weather
6	Juan, Nicky	A can is heated and then "crushed" when added to icy water.	Relationship between temperature and pressure; weather
7	Terrance, Oscar, Laura	A "cloud" of gas forms in a bottle by increasing and then quickly decreasing pressure.	Relationship between temperature and pressure; weather

Note: All preservice teachers' names are pseudonyms.

Observation 1: Preservice teachers developed their ability to use discussion moves to sustain a student-centered discussion.

Preservice teachers appropriated some discussion moves quickly, while others were more difficult. A move preservice teachers used early and often was *revoicing*, which is something they have likely heard teachers do both as students themselves and in their residencies. *Revoicing* can be used to check in with the speaker to see if the teacher is interpreting the idea accurately and can also serve the important purpose of making a student's idea heard by other students. For example, in session 4, Hannah, Keya, and Morgan led a rehearsal in which their "students" were to blow bubbles (with a straw) into a small cup of water dyed with bromothymol blue (BTB), a pH indicator. BTB is blue in neutral and basic solution, and turns yellow in acidic solution. The students noticed that, as they blew bubbles, the liquid turned yellow. One of the students (Bree) observed, "When I stir the liquid, it turns more blue. It goes from yellow to green."

Hannah held up her hand, as the conversation was getting very animated at that point, and said, "Okay, Bree says that stirring the yellow liquid makes it turn green. So, it's getting less yellow, and almost turning back to blue." Hannah realized that for students to take up Bree's idea, she would need to revoice it such that everyone could hear and understand the contribution.

Despite their early enthusiasm for revoicing, we had questions and concerns about the way that preservice teachers used it. For example, they would often take an everyday observation and prematurely attach scientific vocabulary. For example, when a student said, "It's getting more acidic," Keya replied, "Yes, the pH is lower." This modification wasn't necessary at this stage, and in a classroom setting might discourage students from sharing. As noted in chapter 3, vocabulary can become a gatekeeper for participation if teachers focus on language proficiency over sensemaking. Furthermore, revoicing was sometimes used to manipulate students' words so that they fit the teachers' desire for the direction of the conversation. One preservice teacher, typical of this directive approach to revoicing, commented that the teacher should change the meaning of the student's contribution to make the answer more "right" and cause the discussion to "flow" better. The instructors and other preservice teachers, however, pointed out that this sort of teacher-centered practice would likely confuse students and potentially shut down further conversation. As the rehearsals progressed, we noticed preservice teachers' careful attention to revoicing accurately, even if they weren't sure *yet* what to do with the idea.

"Wait time" is another move that preservice teachers picked up quickly, both as teachers and as observers. Similar to the discomfort they felt when there was a lull in the conversation at the touch carts in the exhibition halls in the summer, sitting in silence during a class discussion can be an unnerving experience, so we were encouraged to see the preservice teachers embrace wait time. Grace, in session 3, set a high bar, waiting until everyone had something to add before calling on someone. Several used the "need to see three hands before calling on someone" strategy. They also incorporated "turn and talks" and "take a minute to write or draw" to encourage their students to share their ideas in a low-stakes way before waiting for new contributors to the discussion. As pointed out in chapter 2, for *all* students to take up academically productive forms of discourse, the teacher must scaffold talk opportunities, reinforce norms of inclusion, and seek out all voices in the conversation. We were encouraged by the commitment of the preservice teachers to provide wait time and seek out new voices.

Building upon students' ideas and connecting them to other students' ideas proved to be a more challenging set of discussion moves to master. As chapter 3 points out, though it is critical that there is no single understanding learners must "get to" in a conversation in which ideas are elicited, students should listen to their peers and begin to connect their ideas, such that the group is collectively making sense of the phenomenon. Therefore, though it is important for as many voices as possible to be included, it moves the conversation forward when students can think about what their peers are saying and add on and clarify. Facilitating this type of discussion, as we observed, was a difficult tightrope walk for the preservice teachers.

One example of this challenge occurred during session 4. The teaching team was soliciting working hypotheses from the class to answer the question, Why did the water turn yellow? Students had offered three distinct ideas throughout the course of the discussion: (1) saliva had somehow gotten into the water and was affecting the pH, (2) the water had warmed up and that made it more acidic, and (3) the carbon dioxide from their breath had dissolved in the water and that was making the water acidic. The teachers remained open to new ideas. They continued to deploy excellent wait time. Yet, it was unclear whether the "students" (or the teachers) saw that these were the three primary working hypotheses. They were heading into a long, drawn-out debate about the pH scale when we (the instructors) looked at each other, shrugging, asking with our eyes, "How can they move this conversation forward?"

This instance was the first time that we used a "pause" in the action of a rehearsal. "Pauses" had not been part of our norms up until that point, but we explained that nothing bad had happened in the rehearsal. We both thought it was a great puzzle for us all to think about. Three somewhat related working hypotheses had surfaced—how can we help students talk to each other about the ideas and add on to them (rather than just add more ideas)? We encouraged the teachers to "huddle up" and talk about their strategy going forward; the observers/students also brainstormed about this puzzle.

Following the huddle, the three teachers highlighted these separate, but related ideas; gave each one a name ("the saliva hypothesis, "the heat hypothesis," and "the carbon dioxide hypothesis"); and noted them on the table that was being constructed on the whiteboard. This drew the class's attention to the existing ideas, and prompted them to think about their everyday experiences that might provide additional insight. Understanding that the goal of "eliciting ideas" is not to tie up the discussion with a bow, the teachers paused the talk and let the class know we would (hypothetically) keep investigating these questions the next day. Though helping students connect their ideas in the discussion continued to be a challenge, preservice teachers recalled the "pause and huddle" of session 4 to remind their classmates to relate ideas to one another.

Observation 2: Preservice teachers increased their comfort with letting students wonder, but it remained challenging.

The desire to prove that they understood the explanation for phenomena, as well as all the complicated scientific concepts undergirding the explanation, was strong in our young teaching Jedi. This finding matched observations from the instructors in the Museum Residency, who also noticed that preservice teachers tended to want to display their extensive content knowledge while facilitating at the carts. As observers/students, they squirmed with discomfort when the segment did not end with a final answer. As teachers, they found it nearly intolerable to leave their students hanging. We wondered whether this orientation toward the assignment was hindering their ability to really focus on students' ideas. So, we made structural changes to the assignment to work on this. Though in earlier cohorts we had provided a list of several thinking routines to structure the "elicit" phase, all of the teaching groups had chosen to use Prediction, Observation, Explanation, which, of course, introduces the challenge that the final phase is, indeed, called "Explanation." This was a contradiction to the practice we were trying to rehearse. So we changed the language of the routine to "Predictions, Observations, *Preliminary Explanations*" and pointed out the reason for this shift. At least on a surface level, this had the intended effect—the groups that used this routine all emphasized the tentativeness of the working hypotheses they surfaced. For example, Laura in session 7 said, "Guys, these are just preliminary explanations. We will have lots of time to work on these ideas this week." Also in an effort to head off the "explain it all!" tendency, we strongly suggested particular groups use other routines; for example, See, Think, Wonder tended to generate more shared everyday experiences and questions from students, and the teachers seemed more comfortable with the lack of closure using this routine.

Despite great strides, resisting the urge to jump in with the explanation rather than let students do the intellectual work was an ongoing challenge. Taking this teaching stance is often in direct conflict with what most preservice teachers experienced in their "apprenticeship of observation" discussed in chapter 1. Furthermore, in residency schools, the conception of the textbook as the final authority is a dominant feature. Finally, for many middle and high school science teachers, especially novices, we have found that their scientific expertise is central to their identities. We often observe that they feel compelled to make sure that their colleagues know that they know the science, and letting students wonder seems to call into question their qualifications.

Some preservice teachers are unable to let go of their role as scientific authority in the classroom and feel the need to explain in great depth what is going on. Each year a couple of preservice teachers don't seem to ever be convinced by the "don't tell them yet" approach. Other preservice teachers may see the value of this approach, but remain deeply conflicted, mainly "feeling bad" about students leaving the room "not knowing." They are also concerned about the time it takes to elicit students' ideas in this complex way, as it seems in conflict with getting through the required curriculum. They voiced the time puzzle many times throughout the course—for example, "How can I elicit all these ideas and let us all puzzle about them in a forty-five-minute period? So I want to just pick up on the right ideas and go with those to push the discussion forward."

Practical strategies for dealing with both time limits and uncertainty seemed to assuage the preservice teachers' concerns about leaving students hanging. The debrief in session 7 focused on the ways in which they had practically dealt with this, such as providing "hint" cards to students who needed more scaffolding to make sense of an activity. We acknowledge the challenges around time, and the instructional vision we are asking novices to enact. This does require an understanding on the part of colleagues and administration; we recognize that the work of novices to enact AST also requires a cultural shift in schools—so that teachers are not rushing toward explanations and have time to puzzle through phenomena with a class. We discuss these challenges in chapter 13, but acknowledge their implications in our own work with novices in their settings here in NYC.

Observation 3: Preservice teachers retained the idea that good teaching is something that "just happens," though they learned how to notice things about teaching and to analyze key moments.

Through the work of "eliciting students' ideas about observable phenomena" we hoped to communicate the idea that planning for this type of complex discussion is essential. However, we noticed that many preservice teachers held the idea that discussions "just go how they go"—either the kids are into it or they aren't. To us, they seemed to be saying, "Teaching is just something that happens to me" rather than "I am in charge of good teaching." This incoming stance was one we wanted to focus on because it seemed to negate the utility of intentionally planning for rigorous, equitable discussion; we aimed to help the novices recognize the agency that is inherent in planning.

New teachers may not be aware of the power they have in creating opportunities for students to engage in a lesson. We designed the "eliciting" assignment to help preservice teachers see that "what happens," while not totally predictable, is shaped by teachers' plans. In a more challenging vein, we want our preservice teachers to learn that what the teacher does in the moment also influences the students' participation, as well as the ideas that are available for the class to take up for further discussion. Yet in that learning experience, the practice of "eliciting students' ideas" often bumps up against preservice teachers' ideas about teaching and knowledge. For example, the belief that knowledge is fixed and comes from elsewhere might discourage new teachers from thinking of students as coconstructors of knowledge in the classroom. In this work, we are asking preservice teachers to challenge their thinking about where scientific and classroom knowledge comes from and begin to see students as a source of ideas in science.

In part because the preservice teachers continued to think of *good discussions* as things that happen to *good teachers* (and in part because building norms around offering critical feedback is *hard*), it was challenging to elevate the rigor of the debrief discussion. Focusing the conversation around the "discussion moves" chart helped them discuss instruction without defaulting to "that was good/that was bad." In session 6, Juan and Nicky engaged the class in a discussion of the "can crush" demonstration. From the beginning, they struggled to communicate the goals of the discussion and to keep the conversation focused around the phenomenon. They ended up on a tangent about the ideal gas law, and we looked at each other—where to begin in the debrief? Would their colleagues be able to engage in a discussion about what happened in the lesson and why?

We were pleased when Kate offered this insight: "I wonder if you had a better idea of what the key concepts were and what you wanted the students to predict, you could set that up for them." Here, she highlighted the importance of thinking about what key scientific concepts are needed to explain the phenomenon before the lesson. This comment led to a conversation about how having a better road map for the discussion might have helped. A road map, or plan, would have allowed them to use discussion moves to help the conversation be less about listing all the possible scientific ideas in all their complex depth and more about linking ideas toward sensemaking.

ARE THEY ENACTING THE PRACTICE IN CLASSROOMS?

Despite the challenge of this work, preservice teachers consistently report that they value the time we spend rehearsing and providing critical feedback on one another's teaching. In our final session, a preservice teacher who struggled (in our class and in her classroom placement) with releasing intellectual authority to students reflected that the "eliciting" work had been difficult, but she appreciated the opportunity to be in both the "student" role and the "teacher" role. She admitted that since many of the lessons focused on weather concepts, and she had the least preparation in meteorology, the experience of actually being left wondering was powerful for her. Though she acknowledged it made her uncomfortable, she realized that she was more motivated to learn. She hoped she could eventually learn how to give her own students this experience.

Furthermore, many appropriated the tools from EDU 620 into their own classroom practice, in residency classrooms, during induction, and beyond. An ongoing program of research documents their work on the practice of "eliciting and working with students' ideas"—we are still learning how this work is impacting their practice in the longer term. In the shorter term, we have evidence that they are experimenting with this practice. Preservice teachers and candidates in induction have requested to borrow equipment from the demonstrations so often that we have added these materials to the program "lending library." They use them for formal observations, demo lessons when they are applying for jobs, and even for their edTPA licensing exam. They use the "Prediction, Observation, Preliminary Explanation" strategy and the discussion moves chart in their residency and induction, prompting discussions with clinical faculty and changes to the observation rubric to incorporate a stronger focus on "eliciting and working with students' ideas." Thus, the artifacts from EDU 620 represent important "boundary objects," helping preservice teachers translate their learning from their methods course into classroom practice.

CHAPTER 10

Practice-Embedded Methods Courses for Preservice Teachers

SCOTT MCDONALD, KATHRYN M. BATEMAN,
AND JONATHAN MCCAUSLAND

A foundational idea shared by the teacher education community represented in this book is that learning, including learning to teach, is situated in and shaped by the contexts in which people participate.[1] While we share this conception of learning, the majority of the learning experiences we provide for preservice teachers occur in university settings devoid of the tools, languages, schedules, young people, and other contextual resources present in the secondary school contexts where preservice teachers will teach. This chapter explores the value added to preservice teachers' learning when we bring the varied stakeholders in teacher education together, across institutional boundaries, to do collaborative work in and with the young peopled-places of school. We share our experiences as members of a research–practice partnership with a local school, including its mentor teachers and students, to collaboratively learn, enact, and reflect on Ambitious Science Teaching as it is realized in one particular local context.

The process involves leveraging a partnership more than a decade old to transform a science teaching methods course into an embedded community of practice. The community learning experiences have been organically codesigned over three years by science teacher educators, university supervisors, middle school science teachers, and preservice teachers. We find it important to remind ourselves that all truly ambitious teaching is catered to a particular context in specialized ways. By taking an approach to teacher education grounded in the same notions of learning as the Ambitious Science Teaching practices we advocate, we are able to further our collective understandings of both science teaching practices and science teacher education pedagogies in powerful ways.

THE NEED FOR PRACTICE-BASED APPROACHES

As teacher educators and proponents of ambitious teaching practices, we are committed to the transformation of science teaching and learning in schools. However, the preservice teachers in our preparation programs enter with notions of good teaching deeply rooted in their personal experiences in a variety of school contexts. This challenge has been described as the apprenticeship of observation and is a central challenge of teacher education.[2] Just as learners come to science classrooms with rich and diverse experiences and understandings, so do preservice teachers learning to teach. As others in this volume have argued, we need to design teacher learning environments that do not take deficit perspectives of preservice teachers, even as we recognize that they come with notions of teaching and learning that we are trying to disrupt.

One of the reasons practice-based approaches to teacher education have become more prevalent is that it has become clear that developing preservice teachers' knowledge of new forms of teaching is not adequate to support the development of complex forms of teaching practice. Teacher educators have targeted this problem by offering preservice teachers important approximations of practice that both maintain some complexity and contextual richness while reducing some of the challenges of "real" teaching practices.[3] These approximations of authentic practices situate preservice teachers' learning in activities that help them better envision what practices look like when enacted. Approximations of practice are a central pedagogical approach to providing preservice teachers with visions of possible practices as well as experience enacting them—giving them both reach and footing.

One staple of teacher education is some version of a short rehearsal of a practice, sometimes characterized as microteaching or peer teaching, that asks preservice teachers to teach short segments or entire lessons to small groups of their peers.[4] Recently, rehearsals have expanded the pedagogy of preservice teachers' teaching of each other to include interruptions and integrated discussions that provide more active engagement of the preservice teachers and the teacher educators.[5] In rehearsals, preservice teachers act as learners, and teacher educators interrupt the lesson to have discussions of key moments or specific discourse moves in order to unpack and understand the choices being made. They also allow for "rewinding" the action so that preservice teachers can try different approaches, providing a kind of micro-iteration of practice. Macroteaching, described in chapter 5, is another expansion of microteaching that can provide preservice teachers with a more complete enactment and richer experiences of the variety of practices relevant in different phases of a unit.

Approximations of practices are valuable and effective ways to address the apprenticeship of observation and provide preservice teachers with experience trying unfamiliar practices; however, they are generally situated in university settings with preservice teachers acting as science students. As a result, they are not situated in the authentic context of schools and cannot consider some other challenges in the development of ambitious practices. First and foremost, preservice teachers' translation of that experience to a school context is hampered by the two-worlds problem discussed in chapter 6, where the vision of teaching prevalent in schools is not aligned with Ambitious Science Teaching.[6]

In addition, schools are often places of violence in the form of racism, sexism, heteronormativity, ableism, and other forms of oppression with which our primarily white, heteronormative, able-bodied preservice teachers have limited experience. This adds a layer of complexity to asking preservice teachers to be agents of equity in their placements. The two-worlds problem, grounded in the lack of communication between stakeholders across the two worlds, leaves preservice teachers as the primary, often sole negotiator between the expectations and cultures of their mentor teachers' school, and those of the university faculty. By rethinking how our methods course could provide an authentic situated context for approximations of practice, we hoped to keep the strengths of the supportive university pedagogies while providing a bridge between the two worlds to move the preservice teachers out of the role of negotiator in terms of both pedagogy and equity.

EMBEDDED PRACTICE-BASED METHODS

The decision to move one of our methods courses from a university context into a school emerged organically from an extended collaboration with practicing teachers, some of whom were graduates of Penn State, who eventually became mentor teachers. The relationship between Scott and the group of five middle school science teachers at Springfield Middle School (pseudonym) developed over a fifteen-year period. The activity of the partnership centered around ambitious teaching practices and was supported in the school by a building principal who allowed time for the middle school teachers to engage in their own school-based professional learning community.

In 2017, Scott was frustrated that without access to mentors with ambitious practices, preservice teachers could not see how to make their visions of practice happen—they had reach without footing. At the time, the course had a robust model of peer teaching rehearsals involving multiple rounds of design, teach, and analyze where preservice teachers not only tried out practices, but engaged in substantive discussion of enactment. However, as described in chapter 2, many preservice teachers struggled when they went into their concurrent field placements, saying they were not seeing ambitious forms of practice, and did not feel they could enact what they had learned. So, in the spring of 2017 a group of stakeholders, including Scott, university field supervisors, and the middle school science teachers, redesigned the program to merge the methods course and a field experience, ultimately embedding the integrated experience in Springfield Middle School.

Emergence of the design

The science education program at Penn State involves a sequence of two core methods courses (SCIED 411 and 412) followed by a semester-long student teaching. The first focuses on lesson design and establishes a foundation of shared language for practice, as well as a shared vision and understanding of the nature of the practices themselves with regard to Ambitious Science Teaching and the Next Generation Science Standards. It also includes two clinics where preservice teachers collaboratively develop and iteratively teach lessons to secondary students. Prior to 2017 the second course in the sequence (412) focused on unit-level planning of instruction and an iterative set of three teaching rehearsals preceded and followed by video analysis of teaching practice. It was concurrent with a field placement course (CI 495C), where preservice teachers spent mornings for six weeks observing in schools, and taught a minimum of six lessons over the course of the semester. It was SCIED 412, in combination with CI 495C, that was moved to an embedded model.

In the original configuration, CI 495C required students to be in classrooms every morning for six weeks, from weeks six through eleven of a fifteen-week semester. SCIED 412 was designed to fit into the "empty space" of the semester around those six weeks, and met for three hours on Tuesdays and Thursdays for the first five weeks of the semester, and again for the final four weeks. Thus, the two courses together filled the mornings every day of the semester, which became a critical affordance for the new design. The main structural transformation in 2017 was an amalgamation and reorganization of instruction hours from both courses into one integrated set, and hours were then distributed in a way that made sense based on the academic schedule of the middle school.

Key characteristics of this new design included that preservice teachers would

- be in the middle school for three hours in the morning every day during the first week of school and every other day for the next nine weeks. The school was on an alternating-day block schedule, which meant preservice teachers were there for every instruction day for their students.
- be placed in groups (ranging from two to four students depending on enrollment).
- be treated as coteachers from the first day. Their roles were scaffolded to allow them increasing responsibility over the ten weeks.
- engage in noninstructional activities with mentor teachers and peers, such as designing and grading/evaluating assessments, and short and long-term planning.
- be both codesigners and scribes of the enactment of the units of instruction cotaught in their mentors' classrooms.
- have "class" in a middle school classroom with Scott every day for half of their time.
- be observed by and get feedback from the field supervisor for at least six lessons.
- attend on-campus class meetings one day every other week.

Current time configuration. State College Area School District middle schools operate on a six-day alternating block (A/B) schedule. Therefore, by scheduling preservice teachers for all B days, they would be present for all instruction for a particular class. In addition, B day mornings provided a seventy-two-minute block of instructional time and a seventy-two-minute block of planning time for all five teachers. We divided the preservice teachers into teams and assigned them to a seventh- or eighth-grade mentor teacher's classroom depending on their specialization (physics, chemistry, biology, or earth and space). This design provided half of the morning for working with middle school students, and half devoted to "class" led by Scott in one of the mentor teachers' classrooms during their prep period.

We asked preservice teachers to be present for the entire first week, both A and B days, to immerse them in the school culture and hear what the mentor teachers talked about when establishing norms in their classrooms. Preservice teachers in the first year of the experiment indicated they needed time to process and reflect as an independent community, so we added meetings on campus in weeks when there were only two meetings at the middle school. Thus, the finalized schedule included one full week at the beginning of the semester, every B day for the next nine weeks, and a handful of on-campus meetings. We had not changed the total number of contact hours, but reorganized them.

Mentor teacher team. The building administrator organized the middle school science teachers' schedule to allow extended time every A day for a professional learning community. The team's inquiry stance on their own teaching meant they consistently took advantage of the time to talk, plan, develop tools, and push each other's thinking about supporting all students' learning. They have been identified as models of science teaching in their district, and the Ambitious Science Teaching approach has been adopted by the rest of the district as part of a local effort to align with the Next Generation Science Standards. (Pennsylvania is not currently an adopting state.) Part of the value of embedding the course with this team of teachers was that the preservice teachers could experience the value of being part of a collaborative team of professionals.

Change in experiences

As we transitioned to this new model, we anticipated some of the changes, mostly those based on the affordances this new organization would bring, such as daily access to a group of middle school students and opportunities to engage in a wider variety of teaching practices as a result of having more time in schools. We also had to be responsive to the constraints of the context and needs of a whole new set of stakeholders who had not previously been involved in planning the methods course, including the preservice teachers. The embedded model had its most significant impact on the two pedagogical structures of the original course—the set of teaching rehearsals and the design of a two- to three-week unit of instruction. Over several iterations we developed several new aspects of the course to respond to the new model of instruction: microrehearsals, teacher time-outs, and unit codesign and capture.

Diversity of real middle school students. In prior iterations of the course we did work on students' critical consciousness to help them better understand what it means to teach every child and how their own identity plays a role in their teaching. As we moved to an embedded model, these conversations became more concrete. We could talk about actual students in class, discuss how to support their learning, enact those plans, and then talk about the outcomes. In the course of the three years we have had many examples of conversations resulting from issues brought up by members of the professional community. This past year, one student recognized disparities in how they and their mentor teacher were engaging with girls versus boys. They realized that boys were being positioned as producers of knowledge more often than girls, such that many previously engaged girls became disengaged. This resulted in conversations with their mentor on how to adjust their instruction to be more conscious of gender.

Other conversations centered race. In one example, two preservice teachers were concerned with the sole African American child in their otherwise all-white classroom (teachers included) because, to the teachers, the African American child appeared to be disengaged. This prompted a conversation about how to engage the child more productively, which sparked a change in practice to center the child's questions for serious consideration during a public discussion. While this one move did not solve all their problems, it did open the door to helping the preservice teacher reckon with how their identities, as white teachers, may have mattered to the child's participation and the overall equity of the classroom environment.

We also considered issues of ability when preservice teachers asked about how to support a student diagnosed with social anxiety in a class culture where students were asked to share their thinking in both small groups and whole-class discussion. The value of these conversations lay in the fact that these were not case studies analyzed in class, or incidents that occurred in their field experience where they could talk only with their mentor. The fact that we had a community of more than twenty science educators with different levels of experience and backgrounds, all in the same context and talking with each other during every instructional day, made learning from these conversations a natural part of the context.

Engaging with the full range of practices. One of the limitations of the university-based course is that time did not allow preservice teachers to rehearse all of the different Ambitious Science Teaching practices. Eliciting students' initial models was the most straightforward, as that is always the beginning of a unit of instruction. The hands-on components of most lessons designed to support made it difficult to get to a discussion of ideas in the limited length of rehearsals (twenty to thirty minutes). The most challenging practice was developing evidence-based explanations, as it requires preservice teachers to synthesize across multiple experiences and improve their initial model, which simply does not occur with only three-lesson iterations.

In the embedded model, preservice teachers were coteaching every class with their mentors and were at least partially engaged with the entire scope of at least one complete unit. This meant they were able to see and participate in all of the practices, including planning as a day-to-day responsive activity, not just a one-time act of creation. While mentor teachers talked about planning with each other when the preservice teachers were not in school, they also included preservice teachers in the process and shared their decision-making process with them. This allowed the preservice teachers to see what it means to have a teaching practice that is responsive to students' ideas, not just in the moment through discourse moves, but also at the curriculum level. It helped them to understand the complexity of taking an ambitious approach to teaching and what it meant to improvise and be responsive, while not compromising your core principles.

Microrehearsals. In the new model preservice teachers were in a classroom with students every class meeting, but only one class of students each day. Thus, there was no opportunity for iterations of the same lesson with different students. The preservice teacher now had access to the authentic ideas of middle school science learners, rather than their peers, but the ability to process and learn to respond to those ideas initially seemed limited.

To create iterative opportunities for practice, we developed microrehearsals. By leveraging the fact that multiple teachers were in the classroom, in some cases as many as five or six, we found a way to embed iteration into the preservice teachers' daily experience whenever students were working in small groups. During microrehearsals, preservice teachers paired up to facilitate small-group discussions. One preservice teacher would speak with a group of students while the other observed and took notes around a preidentified aspect or problem of practice. After a short interaction, the two preservice teachers would huddle up, make a plan for what kinds of changes might be productive, and then move to a new group and switch roles. This structure kept the iterative opportunities for learning to talk with students, though it still sacrificed the more extended individual engagement possible with microteaching.

Teacher time-outs. In rehearsals in the university context we could interrupt and rewind during the action to provide opportunities for preservice teachers to discuss or restart a lesson. In the new configuration, within microrehearsals, preservice teachers, mentors, and faculty could have discussion between iterations that did not interrupt the overall flow of the lesson, but there remained times when the coteachers needed to stop the lesson to rethink in more significant ways, often regarding whole-class activity.

The solution developed serendipitously one day as a mentor teacher was in the middle of a lesson where students were attempting to simulate the deflection of different atoms in a magnetic field using spheres of different mass and a large fan. The mentor became increasingly frustrated with the lesson's progression, particularly students' struggle to understand the relationship between force and change in motion in this context. At the time Scott was in the back of the room and the mentor called from the front of the room, asking Scott for his thoughts. This led to a discussion involving the mentor, two preservice teachers, and Scott about the best path forward for the lesson that occurred over the heads of the students. Rather than turning this into an opportunity to distract themselves, the students listened in to the conversation with interest. The discussion included background from the mentor, who explained that students had a unit on force and motion in sixth grade. This led to questioning of the phenomenon used for the unit in sixth grade or the approach to developing students' explanation in that unit. From this description the preservice teachers were able to suggest some potentially productive ways that the two contexts involving force could be linked by the choice of questions they asked the students. In the end, a consensus of preservice teachers, mentor, and teacher educator was reached to not press students at that time about the details of force and instead to add a discussion the following day focused on one of the activities students had used in sixth grade—a cart being pulled by a hanging weight. The lesson moved on, but a precedent had been set that rippled across the practice of all the classrooms. Rather than only huddling in a corner to discuss pedagogical decisions, which still happened, we now had a *teacher time-out*, where the lesson stopped and the teachers in the room had a public conversation about teaching.

Unit codesign and capture. In the original configuration of the course, preservice teachers were asked to develop a unit plan for a ten- to fifteen-day section of curriculum. In the embedded model, we looked for a way to make this assignment authentically connected to the preservice teachers' experience. Coincidentally, as we were planning the course for fall, the school district was forming a committee to engage a cycle of curriculum revision for K–12 science for the school district. Scott and two of the middle school teachers were made a part of the committee, which eventually decided the district would adopt the Next Generation Science Standards. Based on the development work they had already been doing around Ambitious Science Teaching, the middle school teachers were charged with writing curriculum for sixth through eighth grades.

The secondary science department head suggested that the model developed at Springfield guide the curriculum writing, as the Springfield teachers had been developing and using a phenomena-driven curriculum for a number of years. It made sense to bring this authentic task into the structure of the new methods model where the preservice teachers could contribute to this effort. Instead of developing a unit for a hypothetical future classroom, they would codesign and capture the enactment of curriculum at Springfield in lesson plan form that could later be used as part of the districtwide science curriculum. This new process reduced the preservice teachers' practice in developing curriculum from the ground up, but provided them a wealth of new opportunities to think deeply about curricula as a record of teaching. For example, there were multiple teachers at each grade level and, due to the responsive nature of their teaching, their enactments were not identical. This led to questions of how to represent diversity of enactment in a way that would help a teacher enact activities, while seeing that it is not a script. The preservice teachers, as part of the larger professional community, also had to decide what components or artifacts merited inclusion: Should we have samples of student work? If so, how many? Should they be graded and feedback included? Should we have pictures showing the setup of equipment or descriptions, or both? Should we capture students' ideas as they came up in class as a guide to help teachers in the future? What about the questions that teachers asked? The authentic work of curriculum writing led to a rich discussion of the nature of curriculum and its role related to enactment.

Change in outcomes

All I had ever known, from both education classes and being in classrooms all my life, was explicit instruction: a standard lecture on the material, followed by some guided and then independent practice, but in the end all teacher-directed. Every teaching bone in my body wanted to handle the elements of the saltwater circuit kit myself and simply explain how everything worked to our students; I wanted them to know good and well what ions were, and what solutions conducted electricity, before they got anywhere near the materials. That approach, however, would have been completely antithetical to the Ambitious Science Teaching model. (Preservice teacher, 2017)

For this particular student, the challenges to addressing Ambitious Science Teaching started with its introduction in the first methods course, but continued well into the second practice-embedded methods course. They said they understood Ambitious Science Teaching theoretically (they had reach), but could not comprehend how it would ever work in a "real" classroom (they did not have footing). Just as discussed in chapter 9, they were struggling with giving up the role of being an expert and were unclear what their role could be if not to be eliminating misconceptions by providing correct answers. After spending time embedded in a classroom where both they and the mentor used Ambitious Science Teaching as their daily framework, their perspective began to shift, in large part due to affordances of the embedded experience. The preservice teacher learning described in the preceding sections was the result of specific work done on building relationships between stakeholders over an extended period of time.

Our success in helping preservice teachers was and is dependent on our ability to build relationships of trust and a community partnership with a shared vision for science teaching and learning. Building relationships required time for members to be together, which our reconfigured program provided. Relationship building started before the 2017 redesign. As stated previously, Scott's partnership with the Springfield teachers has been ongoing for more than fifteen years in various research projects, professional development experiences, and informal engagements. Scott serves as academic advisor to many of the preservice teachers, meeting with them to discuss their course selection, master's projects, or future plans. Preservice teachers largely move through the final three semesters of the program as a cohort, taking both SCIED 411 and 412 together and then moving into student teaching. Katie and J.D. taught (as lead or coinstructor) all iterations of SCIED 411 during the three years, as well as participating in SCIED 412 during the redesign process, helping preservice teachers to connect with both instructors and each other. This long-term investment also provided preservice teachers with the trust to come to faculty or teachers when something wasn't working, there were concerns with their placements, or they just felt overwhelmed.

The relationships extended beyond the mentor and preservice teachers—the partnership also included the district administration, university preservice supervisors, the student teaching director, and the students. Scott's long-standing relationship with the district allowed the administrative team to trust the science teaching team to experiment in their school. Trust was also required with university administration to disrupt the structure of the typical preservice teacher preparation program. Though this was challenging, it was the relationships fostered over time that provided a fertile environment where stakeholders could creatively problem solve and codesign the best pedagogies to bring preservice teachers into an Ambitious Science Teaching community of practice. It was the relationships that supported the multiple shifts in school culture—preservice teachers working in teams and coteaching with mentors, calling for teacher time-outs, and generally making teaching a public and shared practice.

NOT ALL RAINBOWS AND UNICORNS

While this reorganization had a large number of affordances and addressed a number of key challenges to preservice teachers' learning, we recognize it also had limitations and introduced new challenges. For example, in the previous configuration there was a clear way, via peer teaching rehearsals and video analysis, to support iterative lesson design with feedback. Within the new configuration this was more difficult, as preservice teachers never saw more than one group of students engaged in the same curricular activities. They could see variation across teachers through observations across mentor teams, but they could not see variation across class periods, and also could not teach a lesson once, revise it, and then teach it to another group of students later in the same day. We attempted to address these constraints by changing some of the design features of the learning environment, but they were compromises without the power of the original assignments. This is an ongoing limitation that we are taking into consideration as we engage in new cycles of design research.[7]

We also had not considered the impact of this new configuration on supervisors, and in particular the formal observations they were required to do. As part of their work there was an expectation that they would observe the preservice teachers' teaching and provide them with a formal evaluation. Given that the preservice teachers were spending more time in classrooms than they were before, this would seem to be easier; however, it led to a negotiation about what it meant to "teach" a class. From Scott's point of view the students were teaching every day they were in schools—they were talking with students, eliciting, pressing, and probing their thinking in small groups. Initially the university supervisor did not feel this constituted teaching to an adequate level to allow for formal observation, both because the preservice teachers were not initiating or leading the lesson, and also they were coteaching as part of a team. We negotiated a compromise by segmenting lessons and designating sections to particular preservice teachers as lead teachers of those sections. Of course, this description does not capture the stress and complexity of this process when preservice teachers feel they are in the middle of a disagreement between two groups of faculty who are evaluating them.

This example was part of a larger, ongoing challenge we face that reflects the political realities of working in schools. They are dynamic environments. Stakeholders change, and when they do, so do expectations. Also, mandates come from outside the school that impact our relationships. This is made more complex because the embedded structure opens the methods course up to the influence of so many stakeholders. In the original configuration, Scott had almost complete control over the content and structure of the course, not to mention the ability to preselect the artifacts (student work samples, video of classroom enactment) that were the foundation for professional conversations. This structure and security were largely taken away, and now the "course" was a scaffolded experience in the real world influenced by stakeholders from a larger, more diverse educational community.

SITUATING TEACHER EDUCATION

Teacher education, in the context of our partnership, is ultimately about situating learning in schools where preservice teachers are part of a community of teaching practice. By embedding the methods course within Springfield Middle School, all the individuals who compose the educational context became *real*. Conversations about social justice no longer centered on abstract people who allowed stereotypes to creep into talk of what an African American girl or emergent multilingual, for example, would act like or need. Rather, the nuances of each child could be examined in a shared context. This was one of the multiple ways that, through regular interactions with students, peers, faculty, mentor teachers, supervisors, and administrators, the preservice teachers were able to experience, in a guided and scaffolded way, the career they would be entering.

The embedded nature of SCIED 412 provided preservice teachers with the opportunity to do the work of teachers with the wraparound support of both their mentors and university faculty. At the beginning of the semester, preservice teachers spend most of their time observing their mentor teachers and faculty engage students. They ask questions and during the "class" portion of the day, get a chance to talk about what they are seeing. By the end of the first week, however, as microrehearsals begin, they are active members of the classroom community. From this point, preservice teachers are (re)introduced to the various practices that make up Ambitious Science Teaching, not only through readings and academic discussions, but also through concurrent and collocated field experiences. Through these iterations, they begin to see that each practice varies not just across classrooms, but across individual interactions, and that teaching in this way is not formulaic, but requires them to think responsively. They see, experience, and deconstruct what it means to improvise in a disciplined way using principles to guide decisions. Eventually, these microrehearsals evolve into leading large-group discussions, specific sections of class, and finally teaching entire lessons over a three-day period. Always during this time, the preservice teachers have the support of their peers, mentor teachers, supervisors, and faculty through teacher time-outs and postlesson debriefs.

Realizing the need for time to talk about what the preservice teachers were experiencing, the "class" times during both the embedded meetings and university-based meetings were often spaces where preservice teachers worked through their own questions together. These times were also when preservice teachers, in grade-level teams, would revisit and rewrite the curriculum document describing what they were watching unfold in the classroom. They would share ideas about specific practices and how to best support individual students in their classrooms. They discussed how to frame the introduction of a phenomenon, the way to help students connect ideas across activities, and how to assess students both formatively and summatively. In contrast to the design of curriculum from the original structure of the class, this curriculum was a living, dynamic experience they were trying to capture rather than an academic exercise in envisioning practices they had little experience with in their own science teaching and learning lives.

When preservice teachers were not in classrooms with students, much of their time was spent talking about specific students and supporting their learning. Beginning with an assignment pushing them to reflect on how their identity has been constructed in terms of race, gender, and ability, preservice teachers spent time thinking about the various identities of their students and how those identities matter in their classroom contexts. There were regular conversations about how to alter the learning environment to provide access for all students in robust ways—for example, not only making sure to call on girls more often, but fundamentally changing the way the classroom is structured to address the inequities that often exist between girls and boys in the classroom. Specifically, teachers and preservice teachers gave more time for the class to think independently and discuss in groups before entering a large-group discussion based on the observation that many of the girls struggled to share their ideas in the larger setting unless they had time to think. Even discussion about how students were being positioned in conversations about discipline were interrogated. Often, issues concerning social justice are difficult to address in the moment through microrehearsals or even teacher time-outs; therefore, the longer pause provided by the debrief time in the day was crucial to unpack issues of social justice in the context.

One of the unforeseen benefits of the embedded methods course is the professional development of the mentor teachers. By working with preservice teachers and the university faculty, the mentors were encountering new ideas for their practice, and also gained outside perspectives on their own new ideas. The preservice teachers' beginner status forces an articulation by mentor teachers of practices they now take for granted, thus making those practices more visible and available for the community to unpack and examine. Realizing that some students struggle to participate in the discourse-heavy classroom environment, one preservice teacher looked for ways to support a student with autism in participating in larger class discussions. After extensive observations, conversations with the student, and speaking with others in the school, including special education support staff, the preservice teacher decided to have the student (and others if they wanted) write individual responses to questions on a dry-erase board during larger class discussions. They also planned to establish a classroom norm where partners discuss their writing together before sharing to help the student become more comfortable with speaking. Actions such as these pushed the mentor teachers to change their classrooms and alter their practice in ways that they may not have thought of previously.

As SCIED 412 has evolved over the past three years, both the partnership and methods course have become richer and more integrated. We see evidence of this in how the attitudes of our preservice teachers change from year one to when they enter their field placements for student teaching. Given current interpretations of guidelines on field placements, preservice teachers are placed individually, and thus a maximum of only five students can stay at Springfield for their student teaching. In year one, some of the preservice teachers who remained in the middle school for their student teaching felt they were not being given an equal experience, as unlike the students moving to the high school, they would not be prepared to teach in "regular schools." In year three, some of the preservice teachers placed at the high school asked to be moved back to the middle school, as they felt they had less to learn from "sitting in the back of the class and listening to the teacher lecture" and did not feel they had opportunities to speak with students and hear their ideas. This, for us, represents the power of an embedded methods course because we are providing the preservice teachers not just with knowledge that affords them reach, but also with the footing to fundamentally change the way they think about teaching and their own learning to teach.

CHAPTER 11

Mentoring Core Practices

APRIL LUEHMANN, TODD CAMPBELL, YANG ZHANG,
LAURA RODRIGUEZ, AND LISA LUNDGREN

As university science educators, we seek to support preservice teachers in taking up the Ambitious Science Teaching (AST) core practices; the unique and influential roles mentor teachers play in this pursuit is apparent. Within our courses, before we initiated the partnerships with mentor teachers described in this chapter, preservice teachers were distressed about the inconsistencies between their science teaching methods coursework, where they were expected to engage students with Ambitious Science Teaching high-leverage practices, and their work with mentor teachers in the classroom, where they were asked to take up different instructional frameworks the mentors had found to be useful. This challenge was especially problematic since preservice teachers had not received support in resolving these inconsistencies from either us, as university teacher educators, or their mentor teacher hosts.

Our recognition of the importance of collaborative work with mentor teachers coincided with our realization that, in addition to having years of professional experience, mentors serve as knowers about and liaisons with the community, students, culture, history, and curriculum. As we argue in chapters 1 and 3, research on teacher learning and on equity underscores that everything we do in teaching is only as effective as its ability to meaningfully engage and connect with learners as storied, cultured, historicized, social, and whole people.

In addition, mentor teachers are accountable for their students' science learning and thus serve as important gatekeepers to varying degrees, deciding which learning experiences designed and offered by preservice teachers are allowed in. Further, consistent with chapter 1, we argue that because teachers learn in community and in collaboration with others, engaging preservice teachers in community and in collaborations with us *together with* mentor teachers represents an often untapped and critical feature of teacher professional learning that can and should start in preservice science teacher education.

Thus, we explore connections that can be made across the settings where preservice teachers learn (i.e., university classrooms, local K–12 classrooms), especially in relation to how the preservice teachers can be supported in their enactment of core practices. This exploration includes creating opportunities to collaborate on and negotiate common commitments to teaching and learning and approaches to mentoring with cooperating teachers. Previously we, like other teacher educators, found ourselves somewhat haphazardly placing preservice teachers with mentors based on loose criteria and having little to no interaction or communication with the mentor teachers.[1] Placement criteria could be as superficial as "tenured, high school, biology and willing." Though we aimed to place our preservice teachers with mentors with whom we had longer-term relationships and mutual understandings and commitments about teaching and equity, our needs for mentors regularly exceeded our list of core mentor partners.

In this chapter, we describe the core roles mentors can and do play in our preservice teachers' development so that field experiences can be rich extensions of and complements to the learning goals of our teacher education programs. We argue that mentors' engagement with preservice teachers is an essential aspect of preservice teachers' professional growth, holds potential for mentor teachers' professional development, and informs what we, as a group of collaborating university teacher educators, understand about localized and equitable Ambitious Science Teaching.

We recognize that without shared commitments and teaching practices between their university teacher educators and their mentor teachers, preservice teachers are likely to find themselves enmeshed in a "two-worlds pitfall."[2] As persons with the least power, preservice teachers are left to negotiate the "fit" of expectations from the university teacher educators with those of the mentor teacher with little to no help in finding a balance between these powerful stakeholders' visions or goals. Working closely with mentor teachers and their schools can dramatically reduce the enormous amounts of time and effort that preservice teachers spend in learning to teach, while also juggling expectations of two very different, sometimes conflicting worlds.

Given our concern for preservice teachers as they move between teacher education classrooms and local schools, we describe our efforts to include mentors as integral members of teacher preparation teams, to negotiate localized and equitable visions of Ambitious Science Teaching, and to develop participation structures and mentoring practices that complement the principles of Ambitious Science Teaching. We share some of the approaches and strategies we learned as we worked collaboratively with mentors around these practices to support them in helping their preservice teachers use and locally shape these practices. Pseudonyms are used throughout the chapter to protect the anonymity of participants.

HOW CAN MENTORING AROUND CORE PRACTICES LOOK DIFFERENT?

Mentoring is a core element of teacher preparation programs, playing a significant role in preparing preservice teachers for classroom teaching. The significance of this role can be determined, in large part, by mentors' working assumptions about their roles and positionality, as well as the messages teacher education programs send to mentors about their role. Randi Stanulis and her colleagues have argued that mentoring preservice teachers must involve much more than "cooperating" with the teacher education program to give the preservice teachers a place and responsive support to enact the "what" of teaching.[3] To truly capitalize on the rich potential of field experiences for learning to teach, mentors need to support preservice teachers in understanding the "hows" and "whys" of teaching. Sharon Fieman-Nemser referred to this approach as "educative mentoring," in which mentors colearn with their student teachers as they collaboratively engage in analyses of local practice.[4] Opportunities to "try out, talk about, and then re-tool" Ambitious Science Teaching in classrooms are afforded when preservice teachers are positioned as competent and capable science teachers.[5] As we began to think about the roles and responsibilities of mentors, we also recognized the responsibility of teacher education programs to offer support for mentor teachers, especially with regard to their role in preparing preservice teachers as they work alongside them and with us.

In addition to the assumed roles and responsibilities of mentors as colearners *with* the preservice teachers they are mentoring, we have learned that the *types* of problems they focus on matter. The most productive learning and discussions came when mentor and preservice teachers focused on problems "without ceilings:" those that were open-ended explorations of improving either learning or teaching. Conversations that focused on shared professional problem spaces motivated preservice and mentor teachers to creatively and expansively draw from other resources (people and tools). In so doing, this work fostered new roles and ways of interacting (e.g., more symmetrical collaborations between dyads) and more and richer discussions. This opportunity helped all involved by nurturing shared language, vision, and commitment. These problems without ceilings stood in stark contrast to a focus on judging and "fixing" preservice teachers' skills and competencies, which resulted in increased power differentials and decreased creative and generative discussion.[6]

Problems without ceilings can be further addressed by improving equity and justice in and through high-leverage practices. Mentors have played an important role in preparing preservice teachers for socially just teaching by engaging in and supporting critical conversations and inquiries around power relationships that exist in the classroom, in science, and in society.[7] In Land's study, mentors and preservice teachers used tools like retrospective video analysis and responsive critical discourse analysis to reflect upon their practice and explore how student-student and teacher-student interactions might have been shaped by race and power relationships. Working together, mentors and preservice teachers were able to recognize and appreciate "specific critical, socially just teaching moves."[8]

Additionally, as we focus on mentoring preservice teachers with core practices, we needn't limit the goal to the development of more effective individual preservice teachers; the work done by collaboratively and critically analyzing local implementation of AST can lead to the adaptation and retooling of practices for varied educational contexts. Partnerships are needed "to create systems that co-investigate productive variations of the practices and their associated tools."[9] Core practices are necessarily being further developed and refined in specific enactments of teaching that are tuned to local settings, contributing to the accumulation of knowledge and practices that are supportive of teaching and learning.

What seems most promising about the collaborative work of preservice and mentor teacher dyads with AST sets of high-leverage practices are the inquiry stances these dyads take toward teaching and learning that were not found to be possible unless dyads and teacher educators identified and worked on common teaching practices.[10] Given this, we envision collaborations with mentor teachers whereby Ambitious Science Teaching is a shared commitment in which we understand how collaboration and negotiation in and for local contexts act to shape science learning and support students of all backgrounds to participate. We provide two illustrative examples of how we have been working to promote the development of such collaborations between preservice and mentor teachers across two contexts. We share these as experiences to ponder, not as exemplars to follow.

CASE 1—COLLABORATIVE MENTORING FOR CULTURALLY SUSTAINING AMBITIOUS SCIENCE TEACHING

Get Real! Science, a fifteen-month teacher-education program, is an intentionally scaffolded series of diverse practice-based experiences to develop "culturally sustaining" ambitious science teachers (designed and directed by April Luehmann and Yang Zhang, two of this chapter's authors).[11] Program faculty intentionally prioritize the development of students' home and historic cultures and identities over their science identities and use science to support students' ability to contribute to matters they care about. Field experiences are designed as spaces for preservice teacher agency that span diverse contexts, including those in and out of school, rural and urban, middle and high school, with each of these experiences differently supported by school-based mentors. Preservice teachers are charged with and mentored in understanding, developing, and implementing culturally sustaining Ambitious Science Teaching. Mentor teachers play instrumental roles in helping preservice teachers learn the identities, interests, and experiences of students in a given community, position these understandings as core to all stages of AST, and shape curriculum to meet varied, potentially conflicting goals. The teacher education program offers mentors and preservice teachers events designed to facilitate codevelopment of culturally sustaining instruction that are motivated by and grounded in community needs and interests.

Field experiences in the Get Real! Science program include designing and facilitating a place-based camp in a rural setting, designing and facilitating a two-month afterschool science club in an urban school, conducting field observations in school, and completing two student teaching placements in two different schools. A "system of well-designed practice opportunities" is designed to explicitly consider where preservice teachers are in their professional identity development trajectory and what they need with respect to mentorship.[12] In Get Real! Science, this system of practice opportunities incrementally reduces preservice teachers' access to university support, such as coplanning with peers under professors' guidance, while it increases integration into school-based accountability systems across each of three core phases of the program: (1) summer camp, (2) fall field experiences after school and in school, and (3) spring student teaching. Preservice teachers first engage in significant pedagogical risk-taking in low-stakes settings while considering the role of relevant anchoring phenomena and strategies to nurture student sensemaking through science and engineering practices. Field experiences vary dramatically across the timeline of the program in order to cater to the development of the preservice teachers; the roles of the local mentor teachers reflect these experiences.

Preservice teachers have early opportunities to develop instruction that addresses the four high-leverage practices of Ambitious Science Teaching without facing the intense challenges of situating a unit within a year-long curriculum, addressing a particular set of standards, or preparing students to demonstrate their learning on standardized tests. In these initial field experiences, university teacher educators take on lead roles for mentoring AST while mentor teachers provide connection-making with the local (e.g., local contexts, culture, stories, and histories of the community) and participate as colearners of Ambitious Science Teaching, helping to cater these practices for this particular context. Toward the end of the program during student teaching, school-based mentor teachers assume lead roles of mentoring in their particular contexts while university personnel take on support roles. Across these experiences, the intent is to give preservice teachers safe places that allow them to try and fail, that offer them a high likelihood of experiencing meaningful success, and that nurture their professional identity development as teachers who understand, appreciate, have confidence in, competence with, and commitment to culturally sustaining Ambitious Science Teaching.[13]

To offer a specific example of the role of mentor teachers in Get Real! Science, we next share highlights from summer 2018. Preservice teachers spent their first summer of the program developing and implementing a culturally sustaining AST camp for middle school youth in a particular rural context called Sorrus (pseudonym). This first field experience offered preservice teachers six weeks of place-based science and community networking to plan the five-day camp following an AST design format. On the first day, campers' ideas about local phenomena were elicited to develop a driving question board. In the middle days of camp, campers constructed initial models and engaged in learning experiences to revise thinking. The last days of camp were devoted to preparing and presenting science and engineering explanations at a public showcase. The stages of this first field experience were core to each of the field experiences of the Get Real! Science program: (1) beginning in the community and in the literature, (2) collaborative professional development that involved learning and adapting language and tools, (3) ongoing noticing around implemented instruction, and (4) final whole- and small-group reflections and sensemaking.

Beginning in the community and the literature

University teacher educators and preservice teachers began the program work by using an Ambitious Science Teaching lens to research the rural community's people, priorities, and histories through formal and informal structures. After teacher educators modeled the construction of a driving question board motivated by place-based phenomena in the community (e.g., the soundscape of a rural town called Sorrus—how has it evolved over time?), they led preservice teachers through a number of rounds of inquiry in the places of Sorrus. Each experience added to a whole-group consensus-created checklist of "gotta-haves" for small-group model revisions of the local phenomena. Alongside these science experiences as learners, preservice teachers read and discussed publications about culturally sustaining pedagogy, such as work by Paris and Alim.[14]

Building on these experiences as learners and on other explorations of the local culture, preservice teachers created "Posters of Possibility" of local phenomena that could serve as rich problem spaces for the camp based on importance to the community, access to local experts and other resources, and ideas for empirical tests. These "Posters of Possibility" were hung around a room, with blank posters interspersed for ideas that emerged later. An advisory panel of school board members, students, community professionals, camp directors, and teachers was convened to help two representatives of the preservice teacher cohort discuss ideas and add historical stories, ideas, and details to the effort. Phenomena that were chosen included the reason for the chemical taste of store-bought apples (selective breeding on apple farms), the safety of drinking water from the drainage ditch next to the school (as was the custom for atheletes after sporting events), and the construction of arguments for the protection of recess and the design of the new playground. Preservice teacher teams then worked in teams to develop an original empirical investigation performed in the spaces and communities of Sorrus to address the issues. After these findings were shared with potential future campers (current sixth and seventh graders attending school), preservice teachers used these experiences to develop unit plans for the five-day camp.

Collaborative professional development

Local Sorrus teachers then joined the preservice teacher cohort for collaborative professional development that focused on the first set of Ambitious Science Teaching practices: planning for student engagement with big ideas. Sorrus teachers were introduced to their preservice teacher teams as future classroom mentors. Together they used readings and discussions to critique and improve the camp plans given the goals outlined in Ambitious Science Teaching.

As the group talked through their written plans for the five days of camp, mentors offered feedback and advice. For example, our "Loyal to Soil" team studied the selective breeding of apples in response to a question posed by a young person, "Why does the fruit I get from the store taste chemically?" (see figure 11.1). They planned interviews with soil specialists and conducted an empirical examination studying the soil composition of different fields and taste tests. Feedback that mentor teachers offered their preservice team included insights into Sorrus students (e.g., "We have students with anxiety problems"), information about managing groups, roles and the logistics of field trips, advice about how to handle questions you don't know the answer to, and insight about the science of the local context (e.g., "I don't think you are gonna see much difference between [soil in] this and that field"). Mentor and preservice teachers designed the camp collaboratively, which allowed for similar goals, objectives, and plans. We designed this experience so preservice teachers were positioned to share their plans in educative ways as they explained key elements and fundamental assumptions of culturally sustaining Ambitious Science Teaching to their future mentor teacher while they sought input and advice. Future mentor teachers were positioned as experts of the local culture, camp, and youth. Additionally, they provided their knowledge as they offered feedback and ideas for strengthening the camp plan.

FIGURE 11.1   Selective breeding of apples as a "Poster of Possibility"

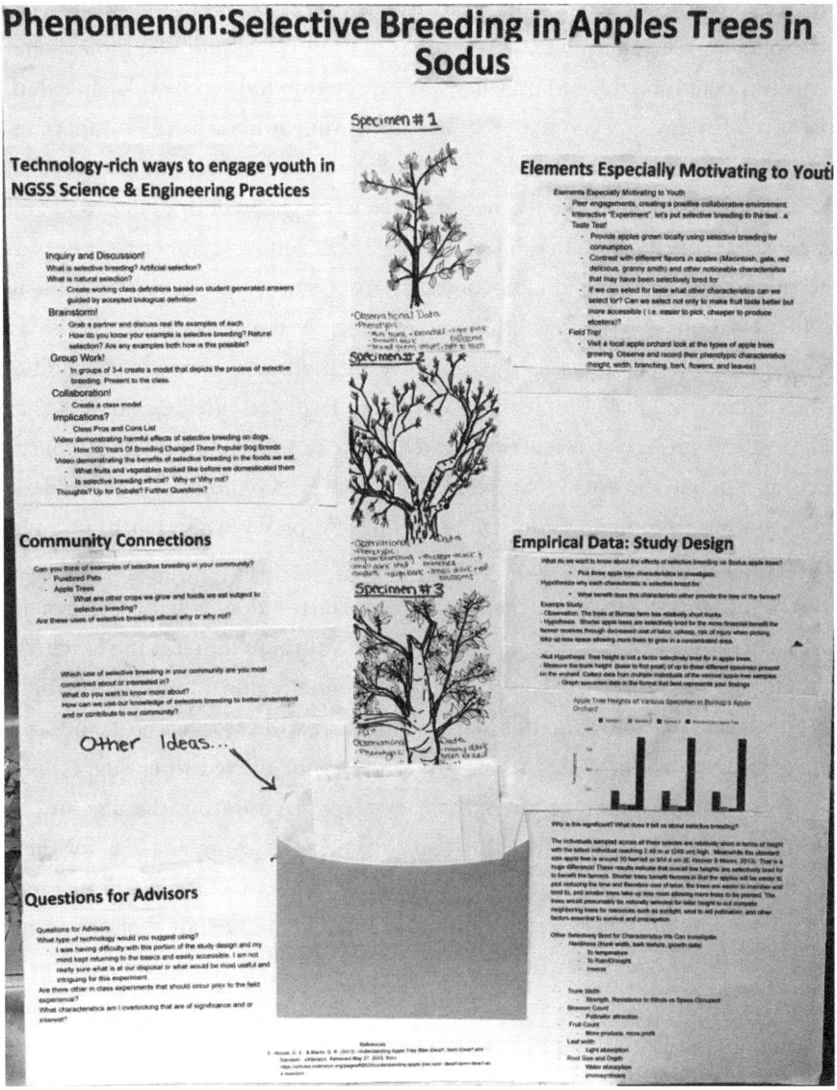

Professional noticing during implementation

Once camp began, preservice teachers led instruction. Mentor teachers were present during instruction to offer advice and provide support with materials management or other background tasks. Before each day's teaching, mentors, preservice teachers and university personnel met to review the day's plans for each of the camp teams. At that time, the university teacher educators asked preservice teachers and mentors to focus their day's observations and teacher time-outs on a given aspect of Ambitious Science Teaching. The professional focus was determined based on what was expected to happen that day in camp. The focus for day 1 was centered on eliciting student ideas as the campers explored information "stations" about each of the three local science phenomena, facilitated by the preservice teachers to learn what local youth understood and cared about. The professional focus for day 2 was equitable discourse moves as preservice teachers planned to coconstruct group norms and expectations. Day 3's focus was on supporting youth's sensemaking given new evidence, as this day was dedicated to place-based empirical investigations and community member expert interviews. Day 4 focused on nurturing evidence-based explanations, as campers and preservice teachers prepared to share their findings and implications at a public showcase that included local press, community stakeholders, other youth, and campers' families the next day. As preservice and mentor teachers enacted AST practices, they asked each other for help during the day's lessons. While the preservice teachers enacted the instruction, either they or mentor teachers were able to call a "teacher time-out" to pause the teaching to reflect, consider options, and engage in collaborative in-the-moment problem solving.

At the end of each camp day, preservice teachers, mentor teachers, and university faculty debriefed as a whole group. Everyone offered three suggestions for improvement and three strengths they recognized during the day, utilizing the Ambitious Science Teaching focus from the morning as a lens for their comments. Mentor teachers often used this debrief opportunity to add rich texture to what might otherwise be more surface-level observations about teaching practice and its impact. For example, mentor teachers shared individual youth's histories of past struggle, disengagement, and detachment, adding significance to observations that all students were enthusiastically participating in a scientific debate. Mentor teachers also contributed insights into local history, family dynamics, learner needs, and community priorities as ways to refine our collective interpretations of the effectiveness of instruction as well as to advise future instructional design. Their deeply rooted perspectives allowed prospective teachers to see classroom interactions in a local light. In the following excerpt, the "Stink Squad" team, studying the invasive species commonly referred to as stink bugs, reflected on the "popsicle stick debate" in which preservice teachers used a cartoon to present a scientific paper to campers that described the possible strategy of introducing a second invasive species to control the growth of stink bugs. Campers were invited to debate the pros and cons, and each camper was given and encouraged to use three popsicle sticks as currency for taking turns and making contributions. A mentor shared this insider feedback to her "Stink Squad" preservice teachers:

> Stink Squad, we had a really good day. More and more, focusing on individual students and building relationships. You with Jimmiqua . . . it was magical. She has stuff going on at home . . . she's off. If you push her, she is oppositional and defiant. So it was great that you said, "For this activity, you can sit away; [but] for the second part, I need you here for this"—she participated in the discussions. She can be difficult; for a few years we've struggled. Another student, Brent, wasn't participating; then he used all three popsicle sticks for discussion. I didn't anticipate that one! Kids that always talk reined in a bit . . . [it] was a safe environment . . . these kids, when they held up their hands and waited their turn, they restated what other kids said. Oh my god! And politely said, "I don't agree with what David said . . ." But . . . where did these manners come from? . . . Brent felt he was safe . . . Everyone was patient, all looked at the person talking . . . the debate got heated, but you had it in such a controlled way . . . The only suggestion I had, don't be afraid to tell kids that they should be cleaning up after themselves. It's a library and it's not our space . . . so just reminding them, push in chairs, pick up papers, hold them to it. When you have classrooms, you'll be glad . . . it's like a hurricane, make them be responsible for their workspaces.

It's clear from these comments that our collective interpretations of camper engagement and participation were importantly informed by and nuanced by the mentor's knowledge of the youth and their lives.

Collaborative reflection and sensemaking

After the showcase on the final day of camp, preservice teachers, mentor teachers, and university teacher educators met together to reflect on the camp. Following the same procedure as daily debriefs, each member of the whole group shared three strengths and three suggestions they identified as core from the camp. Mentor teachers brought their multiple-year histories with the camp and their collaborations with us to bear on these camp-level reflections. Often comments emerged about perceived impacts that far exceeded science learning and participation. One mentor described how the youth in the "ExStream Team" (who studied what youth called "Jesus water" from the drainage ditch and made water stewardship signs for a local pond situated in one camper's backyard) have a newfound appreciation for Sorrus as a town. Before camp, she noted, many youth could not wait to get out of the town. Within a week of camp ending, preservice teachers also prepared individual reflections and collected artifacts to contribute to a team video recording of a multimodal and semistructured reflection, which required them to bring readings and memories of practice into conversations with each other.

The summer camp work described was the first of a sequenced set of field experiences. Common across all core field-based experiences was the prioritization of social justice aims through culturally sustaining Ambitious Science Teaching. In practice, this commitment included continuously learning from communities where mentor teachers were the primary interlocutors to situate science instruction in the social, economic, and historical contexts of local place, and centered learners as sensemakers who used science and engineering to make positive differences.

Thus, each field experience included the four core elements described earlier. Across the core field experiences, there were differences in the levels of accountability and locus of support for preservice teachers. Initially preservice teachers were highly supported by university teacher educators in using local culture and AST practices to enact student-centered teaching with less colearning and facilitation support from local mentors; this support shifted across the fifteen months with less direct intense support from university teacher educators and more significant and central mentoring from local mentor teachers. This shift was important as it honored the development of the preservice teachers and their imminent roles as practicing teachers in established school settings. By the time they reached student teaching, we hoped that they had developed understandings of, appreciation for, confidence with, commitment to, and competence in culturally sustaining high-leverage ambitious practices that collectively empowered them to negotiate the cultural and political dilemmas of advocating for this approach to science education.

CASE 2—PARTNERSHIPS WITH MENTOR TEACHERS SUPPORT ENACTMENT OF CORE PRACTICES

MENTOR TEACHER 1: I think the thing that I like with Todd [university teacher educator and an author of this chapter] being a liaison, he's working with teachers in the trenches, but he's also working with—what do you call? Mentor teachers . . . with those student teachers. He's trying to figure out how do I bridge that gap to what I'm teaching my students here [in university science teaching methods classrooms] and what's actually happening . . . That to me was the ideal person to have in this role.

MENTOR TEACHER 2: Right. Yep. This year I had a student teacher that was going through this with Todd . . . We talked the same language, and so I could—at least it wasn't all Greek to me when he was, "Oh, no. Let's do this and that."

These interview excerpts speak to the value of long-term sustained partnerships with mentor teachers. More specifically, they hint at the promising possible channels of communication that connected preservice teachers, mentor teachers, and university teacher educators. Since 2016, mentor teachers have extensively collaborated with University of Connecticut teacher educators. This work was initiated by the university teacher educators (i.e., Todd Campbell and Laura Rodriguez) with the aim of crafting coherent experiences for preservice teachers across university classrooms and local school settings. In this chapter's second example, we share details related to our motivation for a sustained partnership, how the partnership unfolded with the goal of negotiating and refining common visions and practices in relation to teaching and learning, and how the partnership provided mentor teachers with resources for complementing preservice teachers' early attempts to enact the core practices of Ambitious Science Teaching.

Sustaining partnerships focused on a common vision and AST practices

Early in the 2015–2016 academic year, it became clear that our preservice teachers in science were becoming increasingly frustrated as they entered clinical and student teaching experiences in local schools. Their frustration emerged from the different ways they were encouraged to think about science teaching and learning. The science teaching methods classes were framed by Ambitious Science Teaching core practices, and their clinical experiences involved various other ways of thinking about science teaching and learning. We recognized this as the two-worlds pitfall described earlier in the chapter. Subsequently, we decided to devote increased attention to opening up channels of communication with mentor teachers that supported our joint collaborative efforts to negotiate a common vision of what it meant to teach and learn science.

To accomplish this, we approached a group of in-service teachers who had previously served as mentor teachers and asked them to join us in a long-term effort to learn about the Next Generation Science Standards, while also committing to collaboratively support preservice science teachers. We identified a group of fourteen mentor teachers who agreed to commit time to work with us. Early on, our work focused on developing relationships and trust that we believed would emerge from our group's shared pursuits. We worked to accomplish this by engaging collaboratively in the Next Generation Science Exemplar pathway, a web-based learning environment that focuses on science practices. We found that the focus on this pathway served as an early mechanism for fostering our mutual understanding of the principled commitments of the Next Generation Science Standards.

For instance, one experience we had with the web-based platform illustrates how learning about the Next Generation Science Standards could be seen as a joint experience in which we developed explanatory models of a phenomenon. As part of the Next Generation Science Exemplar experience, mentor teachers and university teacher educators engaged in a multisession "learner hat" experience (i.e., they took on the role of a science learner) where they were supported over several model iterations to explain how to drink from a straw. Over time, as part of the web-based platform, mentor teachers drew on and introduced ideas about how particles move (i.e., kinetic molecular motion) to explain how one of the mentors was able to drink from a straw. Through these joint experiences, mentors were encouraged to reflect on their own experience struggling to explain the phenomenon, and the role modeling played as a sensemaking practice to support them in resolving their difficulty. More specifically, mentors reflected on the functional role that modeling played as a way of working-as-knowing to resolve something uncertain. Modeling was understood as a process of representing ideas with and to others about how things happen in the world and then revising those ideas in light of new evidence or in consideration of new ideas (e.g., authoritative ideas in text or other resources). Additionally, mentors were encouraged to think about the role the teacher might play over time to support student learning in this way. One participant explained how modeling helped her deal with dissonance she experienced related to "giving in to the process that evolves when you stop depending on these tried and true explanations that don't really show deep understanding in a student." The mentor teacher noted that her experience helped her realize that engaging in modeling to explain a phenomenon requires increased levels of thought and connection between ideas that go beyond merely learning about science ideas out of context. Related to the mentor teachers' experiences, we also engaged preservice teachers in similar portions of the Next Generation Science Exemplar pathway as part of their science teaching methods courses. We felt this would further support our early attempts to address concerns about the two-worlds pitfall our preservice teachers had previously highlighted encountering.

Beyond this effort, we oriented our collaborative work with mentor teachers to the Ambitious Science Teaching core practices. We worked with mentor teachers using unit development templates that were framed around the AST core practices that we had previously developed, used, and refined with preservice teachers in the science teaching methods courses. This early work between January 2016 and May 2017, along with continued curriculum design work with AST core practice–framed unit design templates during the 2017–2018 academic year, set the stage for a more nuanced focus on the AST core practices, described in detail in the following section.

Mentor Teachers' Resources

Our most recent collaborative work with mentor teachers during the 2018–2019 academic year focused on approximations of the core practices of Ambitious Science Teaching. In this, we drew heavily on the AST core practices to create templates preservice teachers could use in science teaching methods courses to prepare for and engage in rehearsals of the Ambitious Science Teaching sets of high-leverage practices. Here, templates were supports teachers completed that allowed them to plan for AST instruction (e.g., eliciting intial ideas), with possible sequences of instructional practices they could try out with one another and subsequently use with students in classrooms. An example template can be found in figure 11.2.

We also concurrently used these same templates with the mentor teachers in our monthly meetings, in both preparation for and engagement in rehearsals of the AST sets of high-leverage practices. This consistency allowed us to participate in more nuanced learning about these ambitious practices with both preservice and mentor teachers, with a focus on how AST could support the moment-to-moment sensemaking work of learners in classrooms. Additionally,

FIGURE 11.2   A portion of the Eliciting Student Ideas template

Eliciting Student Ideas Preparation Template

Group Member's Name:

Unit Selected for PT1: [Provide link to unit here]

Part 1.

1. What big ideas are the focus of the unit you've selected? [Big ideas have the power to explain] (List these here):
2. What is the anchoring event for the unit you've selected? [We focus on an event that happens in the world] (Identify this here):
3. What is the essential/driving question that students are trying to answer over the arc of the unit? [What is the "why" or "how" question that will elicit a causal explanation?] (Identify this here):
4. What is the target teacher's explanation for the driving question? [Read carefully.] (Nothing needed here)
5. What is the sequence of events planned for day 1 in the unit that will make up what you will peer teach on Thursday? (Bullet these here):
   -
   -

Note: PT1, above, stands for "Peer Teaching 1."

through these experiences, we refined our approach to rehearsals, while concurrently providing a jointly negotiated vision for supporting preservice teachers in classrooms as they moved from rehearsals in science teaching methods courses, with peers as learners, to engaging students in their mentor teachers' classrooms, where mentor teachers could draw on the rehearsal strategies they learned to support the preservice teachers.

ROLES, RELATIONSHIPS, AND LIMITATIONS: WHAT WE'VE LEARNED

This section highlights what have emerged as core issues in creating localized context-dependent partnerships with mentor teachers. Here we briefly foreground what we believe can be synthesized across the two contexts as the key considerations related to roles, relationships, and limitations, as well as why these are important for supporting such collaborative partnerships.

Roles

Each group of stakeholders played an integral role in the process of learning and implementing Ambitious Science Teaching. One cannot describe the roles of mentors without simultaneously outlining the complementary roles played by the other stakeholders: namely, the teacher educators and the preservice teachers themselves. As highlighted in the description of Get Real! Science, mentors fulfilled their roles as community members who priorized their specific systems, people, needs, adaptations, and impacts for their local contexts. University teacher educators supported both groups of teachers as knowledge and resource brokers, especially as they helped elicit the valuable expertise of mentors. University teacher educators also supported collaborative interpretations of practice by highlighting connections with current academic conversations about teaching and learning, as was seen in the quotes from the mentor teachers shared at the beginning of the description of the second context at the University of Connecticut. Finally, the preservice teachers themselves brought rich backgrounds, expertise, passion, and creativity to the art and science of teaching. Their risk-taking in the varied aspects of teaching provided the focus of learning and reflection for all stakeholders, including, and most importantly, themselves.

Relationships

The risk-taking involved in learning with and from one another requires that trust and relationships be built among the varied stakeholders. This characteristic was emphasized in the description of both contexts. This trust-building may take more time and effort given the likely varied perspectives, expertise, and resources of stakeholders. The two-worlds pitfall discussed earlier highlights the historical issues preservice teachers have encountered as they move across the two important sites in which they find themselves learning to teach (i.e., university science teaching methods classrooms and local schools). In the end, as shared in both contexts, developing relationships mattered, since these provided opportunities to arrive at collective understandings and shared purposes that resulted in productive work toward a common goal, something that can help to increase the permeability between these two worlds preservice teachers encounter.[15] Beyond this, both contexts revealed the importance of recognizing the different roles and expertise those involved in teacher education (i.e., university teacher educators and mentors) bring to this work and how explicit attention is needed to make these roles and areas of expertise a direct part of supporting preservice teacher learning.

Limitations

We very much see the efforts across both contexts as works in progress. In the Get Real! Science context, the camp experience described earlier illuminates some of what was important in strong mentoring relationships and experiences, while omitting many other aspects. Specifically, while local mentors engaged in ways that were manageable and meaningful for them and capitalized on their unique strengths, the depth of the work accomplished around interpreting evidence of learning from student work or carefully critiquing particular practices such as discourse moves was limited by time. In the context of the example of a localized partnership with mentor teachers we have fostered at the University of Connecticut since 2016, there is much work to do in better leveraging mentor teachers as knowers about and liaisons with the community, students, culture, history, and curriculum, as has been a focus in Get Real! Science. Specifically, we see that work is needed in explicitly foregrounding social justice aims through versions of equitable AST like culturally sustaining Ambitious Science Teaching.

CHAPTER 12

Institutional Constraints in Practice-Based Teacher Preparation

RON GRAY AND ERIN MARIE FURTAK

In recent years, we have joined a number of our science teacher educator colleagues as adopters and adapters of Ambitious Science Teaching (AST) tools and resources to support preservice science teacher learning. We write this chapter as teacher educators who have integrated these tools in various ways into our own science methods courses. Over time, we have become increasingly curious about how our own experiences at two different universities—one a high research activity university (Carnegie classification) with science teacher education coordinated out of the College of Natural Sciences, and one a very high research activity university with science teacher education housed in a School of Education—compared to those of our peers. In reflecting on these experiences, we sought to understand how they related to the ways in which AST tools and resources have been used among our colleagues at other science teacher education programs in the United States. These new understandings, we hope, will allow our community of teacher educators to learn from and with each other in ever more engaged ways as we continue to build scholarship through shared language.

We anticipate that AST tools and resources—like other artifacts that organize learning and practice—will shift and change as they move out of the context of their original design and into new institutions. At the same time, we expect that these tools and resources may create or ameliorate tensions within teacher education programs. More specifically, we have wondered about the ways in which institutional constraints in the design of teacher education programs, their associated field placements, and other coursework have challenged these teacher educators in some instances, and afforded innovation in others. This chapter examines what we learned as we looked across a set of teacher educators—many of whom are also contributors to this book—who use AST tools and resources in their own courses, and also explores some of the challenges and constraints that shape their work. We wondered, are our own experiences comparable to theirs? What are common constraints experienced within our community as we seek to prepare preservice science teachers? In a spirit of shared inquiry with our colleagues, we believe these questions will add to our collective learning not only about AST, but about science teacher education more generally.

EXAMINING OUR LEARNING WITH SOCIOCULTURAL THEORY

In this chapter we build on the sociocultural perspectives on learning presented in chapter 1, explore individual teacher education programs enacting AST tools and resources, and examine the ways these tools and resources support preservice teachers as they learn to enact AST practices. We view AST tools and resources as mediating action and collaboration among individuals. However, we acknowledge that these resources and tools are flexible in meaning as they travel across contexts.[1] We begin by identifying the community of AST science teacher educators who use AST tools and resources to achieve the goal of preparing preservice science teachers to enact ambitious practices in their own classrooms. These candidates also work with other members of the community—in particular, mentor teachers in their field placements—as they come to learn these practices.

As we look across the country, however, each of the individual states in which the teacher education programs are located has its own rules and expectations regarding how science teachers are prepared. State departments of education vary widely in their requirements of credit hours or field placement hours prior to licensure, for example. In addition, national consortia such as the UTeach Institute dictate particular courses and course sequences for these programs. Furthermore, national assessments—such as the edTPA, administered by Pearson—require teacher candidates to document and reflect upon their practice in particular ways, and national accreditation programs also impose additional requirements (e.g., NCATE). Ultimately, different individuals in the community—teacher educators teaching different courses, mentor teachers, and so on—play different roles within the preparation of teacher candidates. Given these varying rules and expectations, we anticipated tensions would emerge within these programs as AST tools and resources are applied in service of preservice teacher education.[2]

LEARNING FROM COLLEAGUES USING AST IN THEIR COURSES

While attending an AST conference at Michigan State University in the summer of 2018, we surfaced constraints we had encountered as we enacted AST tools at our own universities. Our own wonderings grew out of these conversations as we realized we had an opportunity to more systematically learn how our colleagues had integrated AST into their own courses and to build a community for shared inquiry around our experiences. While we realize this represents a small sample of our colleagues across the country, we also note the advantage of working with a core group of preservice science teacher educators deeply knowledgeable of and—to varying degrees—beginning to implement AST in their courses.

The eighteen teacher educators in our community represented a range of institutions across the United States, and all described themselves as partially to completely integrating AST tools in their preservice teacher courses. The majority (fourteen) taught at very high research activity universities (Research 1, according to the Carnegie classification system), with fewer (just three) working at high research activity universities (Research 2), and one located at a moderate research activity university (Research 3). Of the groups, the majority (fifteen) taught secondary methods courses and three taught elementary science methods courses. Within these institutions, twelve science teacher education programs were housed in colleges of education with smaller numbers in colleges of science, museums, and other university organizations. The programs ranged widely in terms of level (undergraduate vs. graduate), number of years, and expectations for admission.

While the majority of programs varied in idiosyncratic ways that reflected the universities and states that housed them, three programs shared a common structure. These three teacher education programs—of which ours were two—are identified as UTeach replication sites, or programs modeled after the UTeach STEM Teacher Education Program at the University of Texas at Austin. Following an infusion of funding from Exxon Mobil in 2006, the UTeach Institute formed a partnership with the National Mathematics and Science Initiatives to incentivize teacher education programs across the country to replicate the UTeach model.

To learn more about how our colleagues taught their courses using AST, we collected syllabi from their science methods courses, analyzed interviews with the teacher educators about the role of AST in those courses, and surveyed them to find out specific details about their instructional and programmatic contexts. We requested follow-up information to clarify responses to the survey and interview questions. Because we were curious about how our colleagues taught their courses, we wanted to look at the data we collected from different perspectives. We wondered about when and how often they used certain tools, about the contexts in which they taught, and their perceptions of constraints within their teacher education programs and institutional contexts.

COMMON USES OF AST TOOLS AND RESOURCES ACROSS METHODS COURSES

It was clear that use of AST tools and resources provided common ground in ideas and language across the various teacher education programs. The suite of AST tools and resources—readings, videos, planning guides, and activities for students—predates the 2019 *Ambitious Science Teaching* textbook, and constitutes the set of core practices, primers, and videos developed by Mark Windschitl, Jessica Thompson, Melissa Braaten, and David Stroupe at the University of Washington. (This suite is available on the AST website, ambitiousscienceteaching.org.) At the core are four planning tools that describe each of the four core practices and scaffolds for preservice teachers: *Planning for engagement with big science ideas, Eliciting students' ideas and adapting instruction, Supporting ongoing changes in students' thinking,* and *Drawing together evidence-based explanations.*

Our colleagues also identified additional resources on topics consistent with these core practices of AST, such as managing discourse in the classroom, facilitating effective groupwork, and incorporating modeling in the classroom (see examples in figure 12.1a and b). In addition, the AST website houses a library of videos of teachers enacting AST practices with students in a variety of content areas and grade levels, as well as conversations among teachers and researchers working to enact the AST practices in science classrooms. Taken together, they represent a rich trove of materials to support science teacher educators like ourselves who often struggle to find relevant resources for our methods courses.

This suite of tools and resources formed a common foundation among our group of colleagues, as everyone integrated at least some of the AST tools into their methods courses. The majority (fifteen of eighteen) used all four of the core set of planning tools and, for many of us, these served as the foundation and organizational structure for our methods courses. For others, these planning tools played less of a role and were used as supplements within different organizational structures within the course. For example, one teacher educator used only the "Planning for engagement with big science ideas" tool (a fact that is linked to external factors, as we will elaborate below). Erin, one of the authors of this chapter, was able to use only three of the planning tools, due to the fact that her UTeach-aligned teacher education course at the University of Colorado Boulder serves mathematics, science, and engineering preservice candidates in the same courses.

We also identified additional AST resources, beyond this core suite, that were a common foundation within our community. The majority of us (fifteen of eighteen) integrated at least one additional AST resource, with the most common being the primers—short, research-based readings—which focused on classroom discourse and modeling. Table 12.1 provides a summary of these additional resources. Taken together, this meant that the AST tools, developed by AST scholars for use with their own preservice teachers, have become widespread among other colleagues seeking to integrate AST into their methods courses.

SUPPLEMENTING AST TOOLS AND RESOURCES

Each year, we as teacher educators build or revise our science methods course syllabi, pulling together resources best aligned with our vision of science teaching and the relevant policies at our institutions. Across the field, there has traditionally been little consensus as to the most effective resources from which to draw.[3] This was largely due to our differing visions of effective science teaching and science teacher education. As noted above, however, the eighteen teacher educators

FIGURE 12.1A   Example AST tools

Teaching practice set: *Eliciting students' ideas and adapting instruction*

Overview

Teachers have regular routines that are referred to as *practices*. We have studied how expert teachers work with their students, and we paid attention to moves master teachers make that stimulate student engagement and learning. These educators have been particularly successful in getting quiet and/or marginalized students to regularly participate in reasoning and sharing ideas. All the practices we describe to you here are grounded in research and in studies of the work of experts. You will see that these practices are in many ways unlike traditional forms of instruction.

The practice set we begin with here is *eliciting students' ideas*. This practice is used at the beginning of a unit of instruction. The 3 practices that make up the set are:

1) Eliciting students' ideas
2) Selecting and representing students' ideas publicly
3) Adapting upcoming instruction based on students' ideas

Before we go on, we note here that you have already read (or should read) about the practice called planning for engagement with big science ideas. We assume that you have already organized a unit you plan to teach around a set of big science ideas and selected a compelling anchoring phenomenon that your students will develop evidence-based explanations for.

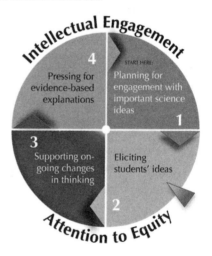

Goals

Your main objective as a science teacher is to change students' thinking over time. So you need to know what your students understand about the core science ideas before launching the unit. The goals of this practice set are to reveal a range of ideas, experiences, and language that students use to talk about the anchoring phenomenon, and to activate their prior knowledge about the phenomenon.

Here is what you are tying to elicit:
- students' *partial understandings* of the target ideas
- students' *alternative conceptions* about the target ideas

FIGURE 12.1B   Example AST tools

AMBITIOUS SCIENCE TEACHING

A Discourse Primer for Science Teachers

T his chapter is an introduction to talk in science classrooms. Talk is a natural activity that we all engage in. As part of our daily lives we use words in various combinations to create speech, and with speech we get work done such as asking questions, providing information, and explaining ideas to one another. But discourse in classrooms can be quite unlike that in our everyday life— there are unfamiliar words that get used, different kinds of work that need to get done with speech (comparing two science explanations, arguing with evidence, critiquing a model, etc.), and rules for participating that aren't always clear. This kind of talk can feel unnatural. Because of this, teachers who want to facilitate productive forms of science discourse with students have to intentionally design opportunities for students to try out new ways with words, and support them in a variety of ways as they learn to "talk science."

For you and your colleagues to experiment with discourse in your classrooms it helps to develop a common vocabulary about talk itself. With just a few basics, you can begin to view your classroom interactions through an entirely different lens. Together you can try out new strategies, debrief them, and advance your practice.

Talk as an opportunity to think: A visit to two classrooms

Here's a very simple kind of logic about talk in classrooms. We start with learning— *learning is a result of thinking.* Certain forms of classroom talk stimulate thinking. Therefore the orchestration of productive discourse in classrooms presents opportunities for students to learn. The key here is discerning productive from unproductive talk. We'll take a look now at two examples of discourse routines, to get a feel for what productive talk is.

Consider a high school laboratory activity that begins with the teacher hanging a mass from a spring-scale at the front of the classroom. The scale reads "1 kilogram." He then produces a large bell jar which he places over the entire scale and attaches the jar to a vacuum pump. "Can anyone share their thinking about what the scale might read if I pump all the air out? Let's take a minute to generate some hypotheses." After a period of quiet thinking, students begin to offer a few thoughts.

Jaden: I'd say it would weigh less—
Teacher: Can you say more about that?
Jaden: Because before you put the jar on top, the air is pushing down on it—the air weighs something, so it's the weight of the thing plus the weight of air.

TABLE 12.1   Use of additional AST resources across teacher educators

AST resources	Number of pre-service teacher educators using this tool or resource
A Discourse Primer for Science Teachers	12
Models and Modeling: An Introduction	8
How to Use Direct (or "Just-in-Time") Instruction in Your Science Classroom	6
Making Changes in Student Thinking Visible Over Time	4
Group Work: Designing for Student Participation	4
Helping Students Talk About Evidence: A Guide for Science Teachers	4
Identifying Big Ideas in Science	3
Card Sort: "Sorting Out" the Big Ideas in a Curriculum—It's a Matter of Priorities	3
Anchoring Events That Can Organize Science Instruction	3
How to Learn from Video: The 7 Basics	3
Scaffolding Students' Written Explanations	3

who attended the conference all used some subset of the existing AST tools in their methods courses (as indicated by their course syllabi). For some, the AST tools and resources made up a significant portion of their courses. For example, David Stroupe and Amelia Gotwals have organized their secondary methods courses around AST tools in a two-semester sequence. The core discourse tools form the backbone of the course, with supplemental AST primers scheduled in a targeted way throughout. Additional readings are at a minimum, with much of the course time taken up by planning and opportunities to teach whole units (a.k.a. "macroteaching").[4] As David explains, "AST is the 'backbone' of the course. AST provides the common language, tools, and practices that preservice teachers need to learn. AST gives the preservice teachers a concrete pedagogy to reify ideas they learn in 'foundations' courses. All of the course assignments are AST tools or framed around AST."

Others use the tools in more targeted ways. For example, Elaine Howes and Anna MacPherson take a comparative approach in their secondary methods courses in a program situated at the American Museum of Natural History. Their students examine a number of theories and conceptual frameworks for teaching, including AST, and are provided opportunities to "develop their [own] theory of instruction by exploring well-established theories of teaching and learning." These include the 5E instructional model (Engage-Explore-Explain-Elaborate-Evaluate), funds of knowledge, and formative assessment probes, among others. As Anna notes, "We focus much more on the 'eliciting' and 'pressing for evidence-based explanations' practices than the other two . . . Even though we work on planning around a big idea and anchoring phenomena, other planning tools are not as central."

As described above, teacher educators across our community supplemented the course to meet their intended goals. As we looked across the syllabi, we counted 252 additional readings that were not a part of the AST tools and resources described above. These varied heavily across the teacher educators, and the majority of these readings occurred in only one syllabus. In fact, it was interesting to note just how varied these syllabi were. While some focused almost entirely on AST tools, others used them as one of many course resources. While some of us supplemented with policy and consensus documents, others heavily integrated research articles and books. We imagine that the variety would only increase if we were to look across science methods courses more broadly and not within this group already committed to AST as a framework.

We were interested in what books consistently appeared on many syllabi aside from AST readings and materials as we hoped they might reveal shared—or perhaps divergent—principles, theories, and ideas among our colleagues. From the syllabi, we counted those readings that occurred on a minimum of three syllabi. There were only nine readings that met this threshold. Looking across these, we noted some interesting patterns, shown in table 12.2. First, and not surprisingly, nearly all of us assigned readings from policy documents such as the *Framework for K–12 Science Teaching* and the *Next Generation Science Standards*.[5] These documents serve to ground preservice teachers in the conversation about science education reform and policy at the national level. Next, the majority of syllabi included additional readings on discourse in science classrooms, although the AST discourse primer (*A Discourse Primer for Science Teachers*) also appeared on many syllabi. Taken together, these resources, which focus heavily upon talk in science classrooms, reveal a strong commitment among our group toward improving classroom discourse, a necessity to teach ambitiously. These readings included the *Talk Science Primer*, *5 Practices for Orchestrating Task-Based Discussions in Science*, and *Ready, Set, SCIENCE!*[6]

In addition, half of us included specific resources to support curriculum design. These included *Understanding by Design* and *5 Practices for Orchestrating Task-Based Discussions in Science*.[7] Next, teacher educators included a number of readings on learning theories, including *How Students Learn*; *Ready, Set, SCIENCE!*; and "What We Call Misconceptions May Be Necessary Stepping-Stones Toward Making Sense of the World," an article that repositions student ideas as assets that they bring to the classroom, rather than as "wrong" ideas that need to be erased and replaced.[8] Finally, three of us assigned *Formative Assessment for Secondary Science Teachers*, a book that provides examples of formats for and approaches to designing formative assessments for the big ideas in science.[9]

We noted that forty-four of the additional readings were about equity and were used by seven of our colleagues. The majority of these resources were assigned by fewer than three teacher educators and thus do not appear in table 12.2. In other words, while seen as important, there was a lack of consensus on equity-oriented readings among our colleagues. For example, Kirsten Mawyer of the University of Hawai'i at Mānoa explained in her interview that the "focus of the secondary program is social justice so there are other frameworks like teaching tolerance that are also integrated into the class." In her courses, Kirsten

TABLE 12.2 AST readings that occurred on at least three of the reviewed syllabi

Category	Count	Examples of resources
Policy documents	16	*A Framework for K-12 Science Education; Next Generation Science Standards*
Discourse	14	*Talk Science Primer; 5 Practices for Orchestrating Task-Based Discussions in Science; Ready, Set, SCIENCE!*
Curriculum design	9	*Understanding by Design; 5 Practices for Orchestrating Task-Based Discussions in Science*
Learning theory	9	*Ready, Set, SCIENCE!; How Students Learn;* "What We Call Misconceptions May Be Necessary Stepping-Stones Toward Making Sense of the World"
Assessment	3	*Formative Assessment for Secondary Science Teachers*

utilized equity-focused resources such as Ladson-Billings's *Culturally Relevant Pedagogy 2.0: a.k.a. the Remix*; Warren et al.'s *Rethinking Diversity in Learning Science: The Logic of Everyday Sense-Making*; and Rodriguez's *What About a Dimension of Engagement, Equity, and Diversity Practices? A Critique of the Next Generation Science Standards*, among others.[10]

Finally, beyond the resources already described, some within our community have created additional tools to adapt AST into their own teacher education programs. For example, Melissa Braaten teaches elementary science methods courses at the University of Colorado Boulder. In her interview, she noted that "[I have] built many of my own tools based on the principles in the official AST tools. Mainly, I make adaptations because the official AST tools don't make sense to elementary candidates in such a short methods course and in the context of their field placements, where science teaching is absent or dramatically different from AST."

Similarly, Doug Larkin of Montclair State University constructed modified AST tools to include both science and mathematics content and examples, as "the existing forms didn't work well in terms of integrating into the STEM teaching methods course." Finally, Ron has worked with Todd Campbell of the University of Connecticut to develop a unit planning template in response to the challenges we witnessed as our secondary methods students struggled to plan instructional units with AST. We see these additional tools as productive adaptations we and our colleagues are making to continually adapt AST to our specific contexts.

THE "TWO-WORLDS PITFALL" IMPACTS OUR USE OF AST

Preservice teachers need opportunities to learn about practice *in practice*; that is, to engage in deliberate practice of reform-oriented pedagogies, such as AST, that they are learning about in their university-based teacher education coursework.[11] Unfortunately, a commonly cited challenge among our colleagues was the mismatch between the AST practices taught in methods classes and those used in the preservice teachers' field placements. Despite a wide variety of practicum and field-placement models, this theme persisted throughout most of our community. These models ranged from preservice teachers placed for a few hours each week in local schools to candidates placed remotely across the nation. In some teacher education programs, as described in chapter 11, the location of the field placements allowed faculty to cultivate relationships between mentor teachers and preservice teachers, whereas other faculty were not able to visit or select mentor teachers at all. In many of these cases, the faculty lamented the struggle to support their preservice teachers in enacting AST practices in classrooms that were more traditional. As Sarah Hagenah of Boise State University put it, "We have to help our candidates as they learn to teach in a traditional setting, but try out reform-based AST practices. This is quite an adventure."

This divide has been identified previously in teacher education as the "two-worlds pitfall," where teacher candidates experience a disconnect between the underlying theories of learning, conceptual frameworks, and practices in their university teacher education coursework and what they encounter in the field.[12] However, this tension of supporting preservice teachers' learning of AST and mentor teachers' often traditional practices was ameliorated, in a small subset of cases, by deep partnerships with mentor teachers who were well-versed in AST, as well as deliberately designed programs that created learning environments consistent with AST principles.

Two examples of these deep partnerships are highlighted in this book. In the previous chapter, April Luehmann provides a second example in the Get Real! Science afterschool program that cultivates opportunities for preservice teachers to practice and realize AST. In this example, preservice teachers are given the opportunity to work within two out-of-school contexts as they build on their experiences from their university-based methods course. Similarly, in the next chapter, Jessica Thompson describes her long-term partnerships at multiple levels of Olympic Public Schools that helped to increase coherence between preservice teachers' experiences at the university and their field experiences. These partnerships are not just with mentor teachers, but also with district science specialists, focused on codeveloping the science curriculum used in partner districts with AST tools.

This kind of "tight coupling" was key to our colleagues' feelings that the practices preservice teachers were learning to enact were coherent with, rather than in tension with, the community and the university. Anna MacPherson noted that when preservice teachers were able to "read about a particular practice, see some exemplars, rehearse with their fellow [preservice teachers], and then try things out with students," greater alignment was achieved. This alignment has also been described as "horizontal expertise" between the university and community organizations that serve as sites for preservice teacher learning.[13]

Unfortunately, building these types of relationships is largely left to us, the teacher educators, leading us to lament the amount of time and resources necessary—but not available—to cultivate the kinds of partnerships that would allow for these types of field experiences. This lack of infrastructure extends to student teaching, which raises concerns across our colleagues about the sustainability of AST practices in environments without support or models. It also suggests we should expand our conceptions of who counts as a teacher educator to include mentors in the field, and involve them more closely and deliberately in the work of teacher preparation.

PEDAGOGIES TO ADDRESS THE "TWO-WORLDS PITFALL"

In spite of the challenges we have identified, the tensions between the universities and field placements can nevertheless serve as important sites for innovation and learning.[14] Given these tensions, many of our colleagues have created structures and routines within their university classrooms to allow preservice teachers to enact AST science teaching practices. We also found many reports of reflecting on videos of classroom practice in university-based courses.

As we describe in this volume, many of us devote time within our university-based courses for preservice teachers to engage in rehearsals or microteaches of lessons, with variety in the ways that these experiences are structured. Microteaching has a long history in teacher education, stemming back to attempts to allow preservice teachers to focus their novice enactments on key teaching practices.[15] These microteaches can be conducted in the field with students, but—perhaps more commonly—the term is also used to refer to practice or rehearsed teaching in the university methods courses with other preservice teachers.[16] University educators usually facilitate the microteaches, using pointed questions as they intermittently pause or interrupt the teaching to allow preservice teachers to reflect on the purposes of what they are doing, to provide feedback, and to allow candidates to even rewind and replay their teaching to better integrate ambitious teaching practices.[17]

In the most common approach for these microteaches, preservice teachers planned a lesson and then selected a small segment of that lesson in which they enacted AST practices in front of their peers and university instructors. Following these rehearsals, preservice teachers reflected on the experience and then, given different university structures, some preservice teachers enacted a revised lesson with students in a field experience.

While some of our colleagues integrated microteaching just once during a term, others integrated as many as three complete cycles of rehearsals in one quarter. A different variation on the microteaching approach has been called "macroteaching" at Michigan State University (as described in chapter 5). This approach was developed to address challenges in having long-term practicum placements at a particular point in the teacher education program. In microteaching, preservice teachers work in a teaching group and complete teaching cycles for a two-week unit that they then teach to the whole class during the semester.

The additional approach of video clubs allowed several of us to focus discussions and reflections in our university-based courses on conversations around AST practices, despite challenges with field placements that were coherent with AST.[18] These clubs feature groups of pre- or in-service teachers watching preselected clips of classroom practice and deliberately framed questions to focus teachers on students' thinking or other elements of classroom practice. As Heather Johnson of Vanderbilt University explained, "I can frame video clubs in a way that requires my candidates to select clips of them engaging in one of the AST practices. We use an AST survey to analyze student thinking."

AST AT UTEACH REPLICATION SITES

The majority of our colleagues feel as though they have a large amount of autonomy in their methods courses. However, those of us who teach within UTeach replication sites are an exception. UTeach is an undergraduate science and mathematics teacher education program at the University of Texas at Austin. Since its inception in 2006, the UTeach Institute has provided funding for universities across the United States to replicate the UTeach model. There are currently UTeach replication programs at forty-four US universities. Inherent in becoming a UTeach replication site is an adherence to a specific program model and to course syllabi, which integrate STEM teacher preparation coursework across a four-year undergraduate degree program. Important to our work on implementing AST in methods courses, those topics commonly included in science methods classes are spread out across multiple courses with multiple instructors serving both science and mathematics preservice teachers.

Three of us teach within UTeach replication sites, including the two authors of this chapter. While all programs are subject to specific standards and policies set at the national, state, and institutional levels, teacher educators at UTeach replication sites also work within a specific programmatic structure that involves combined undergraduate mathematics and science teacher preparation courses in a specific order, as well as specific learning goals and resources for each course.

In our interviews and surveys, all three of us commented on the tension between the UTeach model and AST tools and resources. One tension surfaced around the combination of science and mathematics education students in the courses. Erin teaches mathematics and science together, which requires a more domain-independent approach with AST practices so that they can also be considered in the context of math. She pairs AST videos with a carefully cultivated set of matched mathematics videos that illustrate similar practices. Similarly, the UTeach model suggests use of additional frameworks—like 5E lesson planning models—that may not be compatible with AST.

Finally, we surfaced the challenge of multiple methods courses taught by multiple instructors within the UTeach model. While Ron, for example, may see the benefit of AST in a course, the next instructor may not. Currently in Ron's UTeach program, they have what could only be called a mix of frameworks and resources being used across the courses. Many of the original UTeach-recommended resources are outdated, and coming to consensus on how to replace them is a challenge. This is compounded, as Sarah Hagenah stated, by the "limited amount of teacher education courses," which limits the amount of time dedicated to traditional methods course topics.

Despite the tensions noted here, opportunities nevertheless exist. Even though UTeach sites may not be well represented in the group of teacher educators discussed here, the forty-four programs across the United States have a large impact on science teacher preparation. As this impact grows, it is worth developing a better understanding of the role AST can play in these programs. As researchers, we also see opportunities for research across these sites as they share so much in common.

AST IN CONTEXT: COMMONALITIES AND TENSIONS

In addressing our many wonderings about the community of AST teacher educators who gathered at Michigan State University in June 2019, we have identified areas of commonality and areas of tension. By looking for patterns across the data for eighteen science teacher educators, we found common approaches to adopting and adapting AST tools and preparing candidates to enact AST practices. Taken together, they provided valuable insights into the use of AST tools across contexts for our community.

We have also, however, found that several tensions arise when science teacher educators integrate AST tools and resources into their science methods coursework. We acknowledge that these tensions are inherent in migrating to new contexts, and can be productive sites for innovation and learning to teach science.[19] We surfaced a tension between the AST tools and the "two-worlds pitfall." While having a common language and shared understanding can go far toward ameliorating this challenge, cultivating the relationships necessary within schools and districts remains a time-consuming and challenging task. As we have seen among our teacher educators, successful programs have been built over many years, often integrating the effort with the research agenda of the teacher educator. This may be a necessary step as universities rarely honor this investment of time and energy, especially for those faculty on the tenure track.

Finally, we are aware that our community—which is overrepresented with faculty at research-oriented universities—is not representative of those conducting teacher education at other universities and institutions across the country. This leads us to ask: Who are we missing? To fully understand how AST has grown from its origins at the University of Washington into wider use in teacher education programs across the United States, we need to better understand the tensions that may arise for teacher educators in smaller, teaching-centered universities and nontraditional programs. We also need to better understand the tensions that will exist for teacher educators with research agendas unrelated to AST.

As with all educational initiatives, there are affordances and constraints for teacher educators to integrate AST into their science methods courses. Over the past decade, however, a growing number of teacher educators have taken on the challenge and, as we illustrate in this chapter, they are forming a common commitment and language around sets of core practices as guides for preservice science teachers. In doing so, and partly in response to the need for adjusting AST to their specific contexts, this group has begun to extend AST into a set of shared practices and a common vision that supports our inquiry. We hope the results of our analysis will begin to shed light on the important tensions, both productive and challenging, that we as a community face in this work.

CHAPTER 13

Partnering with Schools and Districts to Improve Ambitious Science Teaching

JESSICA THOMPSON AND DOUGLAS LARKIN

How do teacher educators participate in large-scale educational reforms and contribute to the continual improvement of Ambitious Science Teaching (AST) practices as part of those broader reform efforts? In this chapter, we examine how teacher educators can take part and already are participating in Ambitious Science Teaching educational reform for the purpose of creating equitable and just learning environments for K–12 students. Through partnerships, teacher educators and practitioners in schools can address important questions about practices being enacted, such as: Which practices support learning? Under which conditions? And importantly, for whom? We take seriously the idea that K–12 classrooms are currently not equitable places for all students to learn and that improvement work should focus on culturally and linguistically sustaining practices. We also take seriously the need to do this work in partnerships to support a systems approach to teacher education and to utilize local and relevant data.[1]

In this chapter, we hope to echo many of the themes in this book by showing how Ambitious Science Teaching can be taken up and nurtured through partnerships among teacher educators and practitioners in schools and districts in ways that attend equally to knowledge about equitable science learning and about school reform. It is notable that the current reforms in science education—embodied in the Next Generation Science Standards—require not only new curriculum and assessments, but also substantive teacher professional development and requisite organizational changes and support in order to be enacted. Teacher educators can partner with local schools and districts to build capacity for instructional improvement, and at the heart of the work address improvement of the instructional core—the relationships among students, teachers, and content.

In considering partnerships with local schools and districts, it is worth remembering the insight from improvement science that "variation in organizational context is a core design and development challenge, rather than some externality to be ignored."[2] While some schools and districts may welcome Ambitious Science Teaching or similar instructional reforms at all grade levels, others may have only small pockets of individuals in specific locations who embrace the reform practices and sustain the learning communities necessary to improve and expand their use. School districts and other local education agencies such as charter schools and county vocational schools vary widely in organizational structure personnel, leadership, resources, curriculum, and the characteristics of the students they serve. Variation is a natural, even expected, aspect of professional learning and education reform, and consequently, there is no "one-size-fits-all" approach to science education reform. Yet, coherence within a professional learning community through the distillation of complex ideas into manageable core practices is a worthwhile goal and serves professionalization aims within teaching by valuing knowledge produced by practitioners. To this end, partnerships that identify multiple stakeholders from teacher education programs and local schools is valuable for mapping out problems and improvement aims, as well as for developing structures that support continuous inquiry into the improvement of ambitious and equitable educational outcomes.[3]

Partnerships of this form can be natural extensions of the work we as teacher educators are already doing. Teacher education programs have often made substantial efforts to develop partnerships with local schools and districts. Through supporting preservice teachers, learning partnerships are formed with mentor teachers and school leaders (and perhaps district leaders). But rather than just engaging solely with these individuals, it is important to engage with the educational systems they work in on a daily basis, embracing all of the complexities of the school and district context. As mentioned in the previous two chapters, if we fail to embrace these complexities, we risk working against the very educational outcomes we seek to promote and further contributing to the two-worlds divide between teacher education programs and schools.[4] While well intentioned, many of the current partnerships that rely primarily on interactions with preservice teachers place an undue burden on those teachers as they are expected to bridge learning environments and advocate for their own learning opportunities. To contribute to educational reform and critically examine questions about equitable teaching and learning, it is necessary for partnerships to extend beyond individual classrooms. This means building long-term relationships with stakeholders (including teachers, students, principals, coaches, parents, and community members interfacing with schools); identifying and remaining focused on problems important to school partners; attending to power dynamics among partners; and monitoring the development of infrastructure supporting student, teacher, partner, and systems learning.[5] Such partnerships can build capacity for instructional improvement in terms of addressing the technical, social, and human capital needed to carry out educational reform.[6]

AN EXAMPLE OF DISTRICT-LEVEL CHANGE RESULTING FROM AST PRACTICES

For the past decade, members of the Ambitious Science Teaching team and a midsized culturally and linguistically diverse public school district, which we refer to here as Olympic Public Schools, have engaged in various collaborative projects. Teacher educators began working with preservice teachers and novice teachers who had been placed in the district, then expanded their work to include mentor teachers, and then expanded to partner with school and district leaders and structures such as school-based professional learning communities (some of which included preservice teachers). At this point, the partnership evolved to be a research–practice partnership in which teacher educators/researchers and district-level leaders cowrote grants and developed a longer-term vision and infrastructure for the improvement of science instruction.[7] Over five years, the research–practice partnership brought together teams of teachers, science and emergent bilingual coaches, administrators, district leadership, and teacher educators/university researchers to focus on the improvement of ambitious and equitable teaching practices that supported the Next Generation Science Standards.

Teacher educators in partnership with school district leaders jointly decided to focus on improving teaching and learning of scientific practices (e.g., developing and using scientific models and building evidence-based scientific explanations and arguments) and on scaffolds for the new language demands of the scientific practices, particularly for emergent bilingual students.[8]

While the partnership focused on improvements at the classroom, school, and district levels, this chapter specifically highlights work involving teacher educators/researchers in school-based professional learning communities. Composed of teachers, district-level science and emergent bilingual coaches, science and emergent bilingual teacher educators/researchers, and in some cases school leaders, school-based professional learning communities were sites where improvements to instruction were tested and vetted. The first year of the partnership began with two schools (a middle and high school in the same feeder pattern); we then expanded to nine secondary schools and five elementary schools in the following four years. We designed a professional development model, called Studios, that was embedded within the teachers' work day and allowed professional learning communities to coplan, coteach, and codebrief lessons together.[9] Modeled after Lesson Study, the science Studio model engaged professional learning communities in repeated improvement cycles, whereby teams iterated with specific teaching practices, tools, and practical measures during Studios throughout an academic year.[10] Teams taught the same lesson twice in one day, naming and studying an improvement made to a tool or practice they believed addressed a local and shared problem of practice. During the first few years, professional learning communities met five or six times per year, but eventually we found three times per year was more sustainable for the school district.

Yet professional learning communities did not start from scratch; they used a starter set of science teaching and emergent bilingual teaching practices (see table 13.1) and inquired into how practices and tools supported student participation in disciplinary practices.

Overall the network of professional learning communities developed seven distinct instructional practices between 2012 and 2017. Teacher educators helped name and share these practices across schools. Some of the practices, such as Structured Talk for "How" and "Why" Reasoning and Scaffolding Modeling, spread from one school professional learning community to another, while other practices, such as Science Explanations with Language Functions, remained ob-

TABLE 13.1   Evidence-Based Science and emergent bilingual teaching practices used in the research–practice partnership

Ambitious Science Teaching Practices (Thompson, Windschitl, & Braaten, 2013; Windschitl et al. 2018)	Emergent Bilingual Practices (Lee & Buxton, 2013; Schleppegrell, 2013)
1. Planning for engagement with important science ideas	1. Attending to the development of metalinguistic knowledge, such as language register
2. Eliciting and working with students' ideas	2. Creating multiple opportunities for emergent bilingual students to talk and participate in class activities/discussions with a focus on developing academic language
3. Supporting ongoing changes in students' thinking	
4. Pressing for evidence-based explanations	

Note: We define an instructional practice as recurring work devoted to supporting students' learning and well-being through planning, enactment, and reflection on instruction (Lampert, 2010).

Source: This table was originally published in Jessica Thompson et al., "Launching Networked PLCs: Footholds into Creating and Improving Knowledge of Ambitious and Equitable Teaching Practices in an RPP," *AERA Open* 5, no. 3 (2019), doi:10.1177/2332858419875718.

jects of local investigation. The practices were smaller in grain size than those listed in table 13.1, often attending to turns-of-talk in classrooms. Many practices developed by networked professional learning communities attended to both science and emergent bilingual learning; for instance, Scaffolding Modeling integrated the science practice of modeling with an emergent bilingual practice of supporting comprehensible input with visuals, and the Science Explanations with Language Functions practice focused on composing complex sentences in the context of scientific modeling and explanation.[11] We describe practices that professional learning communities developed and improved as "foothold practices" because they could be implemented daily, addressed issues of equitable and disciplinary participation, contributed to professional learning community alignment by reducing variation across classrooms, and sparked the development of more complex practices across the network of professional learning communities.[12]

Coaches and teacher educators often shared practices across schools, which in some cases was critical for professional learning communities that were struggling to launch joint inquiry work. For example, one of the high school professional learning communities, Washington High School, did not get started in collaborative improvement work during the first year; teachers historically had neither worked together as a department nor partnered with outsiders. The teachers named ideas shared by coaches and teacher educators as "not ours."

This position, in combination with a school-based initiative to focus primarily on reading, made it difficult to get started on improving ideas generated by students, such as the development and revision of scientific models, explanations, and arguments.

It wasn't until the following year, when a science coach and teacher educator shared the Structured Talk for "How" and "Why" Reasoning practice (and classroom discourse protocols) from another school, and the school initiative shifted focus from reading to writing, that the teachers at Washington were able to genuinely see possibilities for their own classrooms. A few teachers began to put the practice into play; then during a Studio with all teachers, the teachers designed a test with the practice to see how teacher modeling might support students in deepening their explanations. Following that Studio, all teachers committed to use the practice and to use an exit ticket developed by another school to understand their students' experiences with the practices. In total they surveyed 1,200 students and then had baseline data to understand improvements as they localized the practice to Washington.

Importantly, tools traveled alongside the foothold practices. Coupling the classroom tools, which were focused on instructional practice, with inquiry tools centered on student data fueled curiosity and innovation, and gave the professional learning communities shared objects for launching and sustaining joint work. Figures 13.1 and 13.2 show examples of tools that were used to support practices and interactions among teachers, coaches, and teacher educators/researchers. Figure 13.1 includes images of a classroom discourse tool the Washington professional learning community modified and tested in year 2, and figure 13.2 is an exit ticket that the network developed to get feedback (data) from students on how the Structured Talk for "How" and "Why" Reasoning practice supported learning.

By the fourth year of the partnership, teachers in the network reported using these specialized tools in 96 percent of all of their interactions with other teachers, suggesting the importance of having tools to help mediate conversations among practitioners and teacher educators/researchers. We had anticipated that tools, as reifications of innovations and a larger vision of ambitious and equitable teaching and learning, might support teacher learning within schools, but we had not imagined that certain tools and practices would provide a foothold into improvement work for other schools in the network.

FIGURE 13.1   Discourse tool for students developed by Washington professional learning community

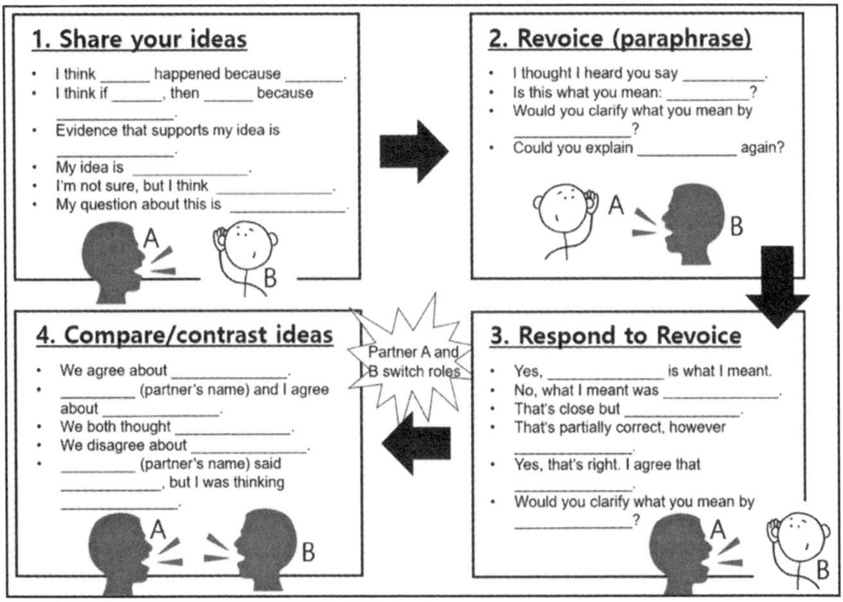

As we observed how foothold practices and classroom tools traveled across classrooms and professional learning communities, we made two conjectures about what supported the uptake and improvement of partnership practices. First, foothold practices and tools aligned teams of teachers and teacher educators to a vision of teaching and learning, helping teams develop common teaching practices that could be tested across classrooms. Second, role actors who spanned multiple settings and levels within the network—such as district coaches and teacher educators who shared the practice of Structured Talk for "How" and "Why" Reasoning—promoted a shared vision of instructional practice and collaboration and contributed to cycles of reification and experimentation. We refer to these individuals as "boundary spanners."[13] We found that they were able to share practices, tools, and a vision of collaborative inquiry across schools, multiplying the connections across schools through their talk.

Social networking analysis helped the partnership attend to the development of these connections as we asked partners to describe connections they

FIGURE 13.2   Student exit tickets providing data for the professional learning communities to investigate specific practices

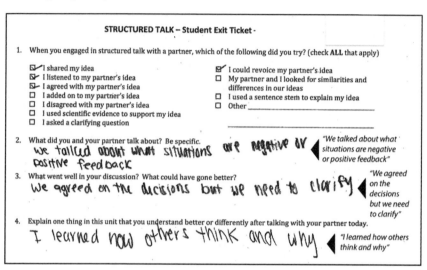

made with others over time and how tools were used in these interactions. The social network diagrams (called sociograms; see figure 13.3) show the number of connections developed over time. Within the first year of the project, teachers (circles), coaches (squares), and teacher educators (stars) were networked as evident by lines connecting the nodes in the sociograms. Over time the number of professional learning communities grew, the network became denser, and teachers began to lead the improvement work in school professional learning communities (narrow triangles in 2016–17).

The partnership project with Olympic Public Schools helped us understand the importance of (1) positioning coaches and teacher educators to promote a shared vision of instructional practice and collaboration, and to contribute to cycles of reification and experimentation with professional learning communities; and (2) leveraging evidence-based practices and tools to develop foothold practices, which aligned people, practices, and tools to a vision of teaching and learning, and reduced variation in inquiries across classrooms. Next, we suggest a set of critical issues teacher educators might consider when developing partnerships oriented toward the improvement of practice.

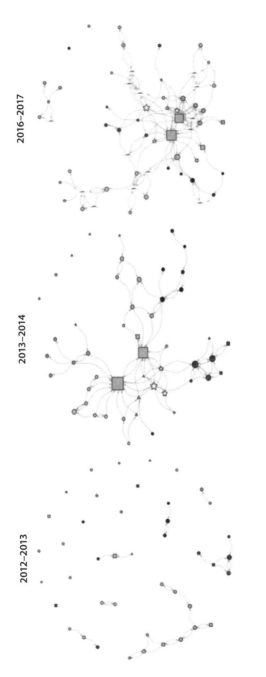

FIGURE 13.3   Social network analysis showing shifts in the network of professional learning communities in Olympic Public Schools over time

2012–2013

2013–2014

2016–2017

CRITICAL ISSUES IN DEVELOPING PARTNERSHIPS TO IMPROVE PRACTICES

A number of critical issues may impact the implementation and improvement of core practices in a school or district, and attending to these is a key component of a systems approach to fostering the type of pedagogical change envisioned within Ambitious Science Teaching and similar reform efforts. The history of instructional reform in science education suggests a continuum of efforts that range from large-scale and detailed changes in curriculum materials and pedagogy to more modest attempts to redirect the nature of classroom instruction at the unit level of study, to small-scale reforms in curriculum materials and practices that reflect contemporary standards and understandings about student learning. Each approach requires some reckoning with gaps between envisioned reforms and the actual implementation in classrooms. In the case of the partnership between the Ambitious Science Teaching group and Olympic Public Schools described above, one salient feature is that those envisioning the reforms and those engaging in the implementation were in close partnership for the duration of the effort, a key feature of research–practice partnerships.

Stepping back, there are several critical issues to consider as partners begin engaging in the design of professional development and the implementation of efforts to improve science instruction. Here we draw on the widely used Loucks-Horsley et al. framework for the design of professional development for teachers of science and mathematics to help articulate some of the issues: building capacity for sustainability, making time for professional development, developing leadership, scaling up, ensuring equity, building a professional learning culture, and garnering public support.[14] In the interest of aiding readers of this chapter who wish to develop partnerships between teacher education programs and districts, we pose each of these concerns as a question that a group working on the issue might ask themselves.

How are we building capacity for partnership sustainability?

Implementing Ambitious Science Teaching and approaches to improving instructional practices in a school or district is a long-term project, and it is worthwhile to consider from the outset how this effort will be sustained over time. An important aspect of building capacity is thinking through how any given initiative will fit into the roles and responsibilities of individuals in their existing work. In the broadest sense, the notion of building capacity means ensuring that there is both space and time for people to do their current jobs as well as take on new responsibilities. Making the decision to prioritize a particular project, such as the implementation of AST in a district, necessitates having honest conversations within and across partnerships about how this will impact the existing allocations of time, attention, and effort. If the new initiative simply adds to the workload of an already overwhelmed partner, it is less likely that it will get the time, attention, and effort it deserves.

True partnership work requires that lead stakeholders—in this case, from both the university and the district—have a seat at the table for decision-making. Part of building a robust and sustainable structure is identifying the key contributors who will work across the partnership, such as coaches and teacher leaders. Establishing common collaborative learning structures that are workable within the district is equally important, as is setting expectations for communication. For example, how often might there be a regularly scheduled check-in or meeting? What are the expectations for planning something like Studios (described above)? What promises are stakeholders making to each other? Being explicit about these expectations may feel awkward at first, but this may also prevent misunderstandings later. Further, making clear plans for the systematic collection of data with an equity focus will help stakeholders evaluate the work and make necessary improvements.[15] As shown in the earlier case with the Olympic Public Schools partnership, recognizing a need for district-level coaches and finding a way to hire them into the project was an important first step. Over time the coaches became indispensable influencers, and a key component of sustaining the project beyond the life of the grant.

Are we budgeting adequate time for teacher learning over the year?

We know from the research that high-quality professional development efforts are more likely to bear results when sustained over long periods of time and integrated into the daily work of teachers rather than being wedged into already full days of teaching. In the Ambitious Science Teaching and Olympic Public Schools partnership described above, professional learning community time had to be periodically adjusted to support long-term sustainability. The first two years, professional learning communities engaged in five or six Studios per year. As the district began to pay for Studios, this number of full-day releases was not feasible. Instead the partnership reallocated time already devoted to professional learning and designed new structures for professional learning communities to examine student work and continue the focus on improving practices during this time. This sort of infrastructure adjustment made use of existing structures but reimagined how partners could provide continuous learning opportunities.[16]

What leadership roles and structures are we building into the partnership?

As should be clear from the example above, the development of a professional learning community around AST is hardly a "turnkey" type of process. For such a community to be sustained and grow over time, certain individuals serve as powerful influencers in the social network of this community. While some of these nodes have formal leadership positions in the district, or are explicitly identified as coaches, others may simply be invested in ensuring that knowledge and practices move through the network as a whole. In building a partnership, it is worth thinking about what roles are envisioned for different "nodes" and how different individuals need to be supported in the school or district to create a culture of improvement.

New teachers often play a critical role in the network, and may be powerful influencers through their desire to improve their teaching and ask questions of people in other nodes in the network who help drive and sustain the network as a whole. While new teachers may in fact take up leadership roles and be Ambitious Science Teaching influencers, there is also a two-way flow of information between any two nodes in the network, and this impacts new teachers in their socialization into the district in critical ways. We prefer to view the role of new teachers as nodes who can both influence and be influenced, rather than simply as the standard-bearers of change (which may be an unrealistic and unfair burden to place solely upon novice teachers). Rather, we view the partnership as a way to support leadership practices that spread the burden and responsibility for making change across the network of novice and experienced teachers alike.

How do we scale up AST practice improvement in a district?

The question of scaling is very dependent upon the structures and leadership roles built into the partnership. A nimble partnership can make meaningful changes during an improvement cycle—such as the hiring of district-level coaches or expanding to new sites. When attention is paid to emerging influencers within the district, this flexibility allows for such influencers to be important resources for teachers who are looking for ways to begin using the core practices (or in the case above, "foothold practices") in their own classrooms, and thus participate more fully in the next iteration of reform in the district. As the Ambitious Science Teaching–Olympic Public Schools partnership expanded to more schools, school-based professional learning communities selected a foothold practice as a starting place. Coaches and teacher educators helped professional learning communities consider local problems of practice and then shared the foothold practice as a "change package."[17] Change packages included a prototypical sequence of the practice, a description of the underlying theory of student learning, classroom tools developed by other teams, exit tickets for professional learning communities to collect data from students on how the practice supports student learning, and case studies with video of professional learning communities inquiring into the foothold practice. In this way, the practice was made concrete and could be easily shared with other schools as a form of technical capital. Importantly, the emphasis in the change packages was on identifying local problems of practice and on collective inquiry that allowed teachers to take ownership of the improvement work. Moreover, by having common practices across schools, "influencers" could accumulate and share knowledge across schools and grade levels about those practices.

Investing in discussions early in the process between all members of the partnership will help to shape the overall uptake of practices throughout a district. Early decisions in this effort include how many teachers should be involved, whether the partnership includes only those working with student teachers or is cast more widely, and what grade levels and schools will be targeted for professional development first. Design-based implementation research using recursive design cycles, in which new ideas driven by partnership questions are tried, assessed, and modified in (relatively) rapid succession, can be an important tool in understanding what works and why in particular contexts.[18]

How do we keep equity at the center of the work?

There will be situations where the equity goals of Ambitious Science Teaching are in tension with existing structures in the district and/or the university. One way to keep equity at the center of the work is to explicitly build measurable equity goals into the project itself. If there is an external evaluator or other accountability mechanism within the partnership, ongoing questions about the equity goals can prevent equity from being swamped by all of the other pressing concerns, and help maintain focus on what the group has already deemed important. In the Olympic Public Schools case discussed above, equity goals were woven into the project as a whole, with the outcomes of the emergent bilingual students a key indicator of success.

These discussions about equity are important to have early on in the process, so that there are clear expectations about what Ambitious Science Teaching will and will not be able to accomplish during a given time frame. For example, in a district with a long history of academic tracking, the conversations about equity are going to be very different from those in a district with a deep commitment to inclusive classrooms and differentiated instruction.

One common occurrence in partnership work concerns how critiques are communicated. It is all too easy for one side of the partnership to feel that a particular equity critique is unfair because they are doing the best they possibly can with the given constraints. For example, a university faculty member may point out that adequate resources are not being provided to English language learners in a classroom, or a school partner may call a suggestion for grouping strategies from a university partner as "unrealistic." Such critiques can easily lead to strained relationships, especially if norms in communication have not been established. Indeed, such comments risk being interpreted as personal attacks. In both instances, the underlying equity issues (dare we say phenomena) may be masked by the critique, and a more helpful approach would be to identify the specific issue at hand in language that does not close off possibilities for a solution or imply that ownership of the issue lies solely on one side of the partnership.

What is our vision for a professional learning culture, and does our planning reflect that vision?

It is worth drawing a contrast between professional development as commonly practiced in the United States—as interruptions or additions to the school day—and the practice of lesson study, or *kenkyuu jugyou*, as practiced in Japan.[19] In Japanese schools, there is a strong culture of professional learning with a focus on teaching as part of teachers' daily work. Time is allocated for teacher coplanning and analysis of teacher observations, which allows for teachers to focus their attention on understanding how and why the observed lessons address learning goals and promote understanding.

In building partnerships to foster the use of Ambitious Science Teaching, it is worth taking stock of the school's or district's operational definition of professional learning. In districts where teachers have little time for common planning, educators may have difficult discussions about the need for opportunities to develop a professional learning culture (as well as issues of power and agency in the school, responsibility of the university partner to be a resource in that professional learning culture, etc). It is possible to build such structures into American schools, as in the case of Studios in the Ambitious Science Teaching–Olympic Public Schools partnership, but there must be explicit conversations about the vision of teacher learning given the limited structures and time available to be devoted to professional development in schools. These conversations need to include debate about instructional practices and underlying theories of student learning, as well as discussion about what counts as meaningful data to teachers.

Additionally, as part of this professional learning culture, it is important to detail the needs of the school or district, and develop clear strategies for identifying and communicating problems back to the leadership team. One principle of improvement science is that there is an expectation that things will not always go as planned, and that existing tools, programs, and processes may fail upon initial implementation in a new context. Consequently, it is important for the design to ensure that such failure happens early in the implementation, so that lessons may be learned quickly and there can be a rapid response. When teams "fail fast" in this way, they carry forward these lessons to the redesign of the next iteration of the project. Such an approach is possible only when the design of a given project includes both the expectation that problems will occur and the belief that there is a way for them to be identified, discussed, and addressed in a timely manner.

In what ways are we garnering public support for our efforts?

Even in schools where science education reforms that embody the ideas of NGSS have taken root and produced successful student outcomes, teachers and administrators continually find themselves having to defend unfamiliar pedagogical approaches to parents and community members.[20] Having allies who understand the powerful learning that can take place in teachers and students alike with AST is crucial for sustaining a professional learning culture. In the case of the Ambitious Science Teaching–Olympic Public Schools partnership, it was important to meet with district leaders and principals early to gain their support.

Yet it was also important to develop structures that supported their learning in particular. We held annual meetings, shared data, and invited principals to attend Studios, but their involvement in the in-depth understanding of teacher and student learning remained minimal. In the fourth year we added a learning-walk component, where teacher educators and coaches joined principals as they dropped into classrooms and conducted brief observations. Tools and protocols were developed, but importantly, we were able to use an existing structure to engage in conversations about everyday teaching and learning and the vision of the educational reform.

CONCLUSION

It is important to recognize that the vision of Ambitious Science Teaching presented throughout this book is not just a curricular reform, but a bold transformation of science teaching and learning. Such a reform will require changes in how teachers teach, what curriculum they use, what assessments they value, how instructional time is used, and what teaching and learning look like in the classroom. The enactment of this new vision of science teaching is more than just a needed shift to curriculum and pedagogy, and framing it as such underestimates what is being asked of teachers, students, and schools and of us as teacher educators. Rather, we suggest that it is preferable to honestly assess the magnitude of the task of science education reform in a given system, and begin where efforts can be leveraged over time to build greater capacity for improvement.

A Community of Teacher Education Scholarship

SCOTT MCDONALD, KAREN HAMMERNESS, AND DAVID STROUPE

This book contains the collected stories of a growing community of science teacher education scholars working to reframe science teaching and learning in schools. This effort is meant to disrupt schools as they currently exist, and to work collaboratively with colleagues in formal and informal spaces to remake science learning environments with a focus on equity and justice. Our community of science teacher educators is nascent, yet is beginning to strengthen ties to the larger existing community of teacher education. These efforts, in turn, better support a collective, expansive engagement in and inquiry into this transformational work. Our vision and work also recognize the culturally complex ways of seeing and acting in the world as it is. These cultural ways of seeing and acting inform a professional pedagogical vision—how we choose to focus our attention, how we make meaning in our contexts, and the way we make representations to structure and share those understandings. We have chosen to reify our vision in the form of core practices and a set of principles of critical consciousness and to ground the vision in stories of creating this new reality.

The articulation of core practices and principles is our way of externalizing our thinking about teaching and teacher education so it can be put to critical examination—both by us, other scholars, and practitioners, and by the students and communities we ultimately serve. Articulating a model is a central practice we ask science students to engage in when they begin their investigations of scientific phenomena. The phenomena of teacher education are more complex system-level problems involving multiple stakeholders and institutions. Teacher education is a three-story, nested problem—teacher educators are designing learning environments for preservice teachers who are learning to design classroom environments for students who are learning science.[1] As a result, our models need to consider not only the goals we have for students in science classrooms, but also how those goals are intertwined as drivers of our goals for preservice teachers. In addition, our work occurs in a large variety of institutional and social contexts, including schools, museums, universities, and communities. We see this diversity of context as a strength of our community that can allow us to better understand our thinking, and provide a rich context for testing our model(s).

The three interacting layers of teacher education can be seen as analogous forms of investigation and model building, simply differentiated by the community engaged in the investigation and the phenomena of interest being modeled. For students, the goal (currently defined by the NGSS) is to develop explanatory models of natural phenomena that are grounded in their experiences with the natural world, ideally emerging from their lived experiences or issues in their communities, and that are more productive than their current understandings of those phenomena. For preservice teachers, our goal is to develop a set of practices and principles that serve as a kind of explanatory model for the phenomena around facilitation of student learning. Just as with students' science models, preservice teachers' models of teaching should be dynamic, contingent, and open to regular testing against the criteria of supporting all students' learning in science classrooms. For teacher educators, our goal is the development of an articulated set of core pedagogies and principles that can serve as a model for the support and facilitation of teacher learning. Just like the models on the other two layers of the system, these pedagogies are contingent and coconstructed with science teacher educators and preservice teachers in a community of inquiry. Also like the other two layers, the models should be tested, questioned, studied, developed, and iterated over time. In all three layers, externalization of thinking for critical examination by the community is a foundational activity. It is also only the starting point for the work, as the models are never complete. Our teacher education model in the form of principles and practices is intended to lead to equitable and just forms of teaching where students are engaged in doing the work of science. Our stories in this book are the beginning of the work of externalizing examples of the model in practice that describe a set of teacher education pedagogies that support preservice teacher learning of science teaching. To carry this work forward we must be willing to turn a critical eye on our own models, constantly measure them against our ideals, and communicate the results of our process.

As part of developing the critical consciousness that we want in our preservice teachers, we must ground our practices in the ideas and identities of our preservice teachers and recognize and support their role in codesigning our teacher education pedagogies. If the system of practices and pedagogies is not dynamic, contingent, responsive, and fundamentally equitable and just, then it will not succeed in disrupting the systems of schooling and society that oppress and do violence to students. We must focus simultaneously on our own examination of the role of our identity in teaching while also structuring preservice teachers' experiences for unpacking their own power and privilege. Across the chapters, readers can see different approaches to the equitable structuring of teaching that are attempts to experiment with forms of practice or principles that can expand our teacher education repertoires—for example, exploring translanguaging to support emerging bilingual science students, expanding and rethinking the role of and relationship to the mentor teachers or university supervisors, rethinking how to structure or reimagine rehearsals of teaching practices, or flipping the script on what elementary preservice teachers bring to teaching science. Clear articulations and descriptions of practice allow our community of inquiry to grow and modify both teacher education pedagogies and the teaching practices they are meant to engender. They can also lay the groundwork for self-study of the impact of our teacher education pedagogies, both on preservice teachers' teaching practices and on students' science learning.

Lortie once described the culture of a school as an egg crate, with each teacher living in a separate world, existing in parallel pedagogical lives.[2] Teacher education is a more extreme version of this problem, as we do not even share the context of the same school. Even the authors of this book are scattered across the United States, from Hawaii to Connecticut and from Seattle to Greensboro. Communication across geographically separated programs makes community and collaboration challenging. In addition, historic avenues for teacher educators to learn from each other, such as sharing syllabi from methods courses and chatting at conferences, do not unpack and expose the contextual factors that shape our pedagogies and both inform and constrain our choices, nor do they articulate the underlying purposes of tasks and tools. Syllabi also fail to capture the complexity of teacher educators' enactment, which we know from research and our work in teaching to be the most important factor in learning. To break down the culture of separation, we are working to build a kind of invisible college of teacher education that leverages the shared language, tools, and practices we are developing to open rich lines of communication and inquiry between our sites. This book is a representation of that sharing, and its creation has been a powerful learning experience for the editors and authors, as it has been an intimate and detailed way for us to learn about each other's work. We recognize that learning is situated, and our shared vision for teaching is shaped locally; we also believe the diversity of contexts and the resulting variation in our local pedagogies and practices provide a remarkable opportunity for us to develop a knowledge base about teacher education.

PRINCIPLES REDUX

In the introduction we articulated some principles that we felt could orient the reader to the rest of the book and illuminate particular strands of connection across the chapters. We return to those principles now to help contextualize where we see opportunities for our growth as a community of scholars.

Principle 1. Teaching in the complex and diverse schools of today is not a natural activity, but it is learnable. Preservice teachers must have structured and supportive learning environments if we want them to develop the complex and responsive practices that will lead to equity and justice in schools. Teacher educators must articulate and open to scrutiny the core practices and principles we have as outcomes for preservice teachers and the pedagogies we use to accomplish those outcomes if we want to make a case for our value and impact.

Principle 2. We view learning at all levels of the teacher education system as situated, so just as we drive teaching and learning for science students with their ideas and experiences, we need to drive science teacher education learning with preservice teachers' ideas and experiences. In order to develop productive learning environments for preservice teachers, we must design our pedagogies with their agency and identities in mind. We need preservice teachers to engage with developing models of teaching in the same way science students engage in developing their models of phenomena—being open to improvement through investigation and testing.

Principles 3–5. As a community of teacher educators we take an inquiry stance in teacher education, exemplified in the stories of this book, where we open our practice to examination and talk about our work in contingent and curious terms. This shared inquiry is made possible by our shared set of language, practices, and tools. Articulation of our models and our work as a community toward reaching consensus on initial models is a critical part of our learning process, and while it requires reification, it is not meant to be canonization of the models. We are not trying to create scripts that deprofessionalize teachers and teacher educators and constrain them with "best" practices. The purpose of our work is to develop and clarify a core set of language, practices, and tools, both for ourselves and for our preservice teachers.

Principle 6. All of the work we do around core practices needs to also be grounded in developing critical consciousness in ourselves and our preservice teachers. Our community is committed to the disruption of a system of schooling that is racist, sexist, homophobic, ableist, and unjust. We cannot be satisfied with being allies and advocates who only point the way toward equity; we need to be stewards and accomplices who are engaged in the work ourselves. We need to engage in our own journeys of self-examination if we hope to ask our preservice teachers to both develop equitable practices and be change agents with their future teacher colleagues.

THE FUTURE

The contributors to this book, like any group of people who share a community, are part of a complex social network that has been developing for decades. There are colleagues from the same institutions, and authors who were students of other authors. We have published work together, written proposals together, and had joyous, commiserating, and bonding social time together at conferences and retreats. Many of us gathered in East Lansing, Michigan, in June of 2018 to talk about Ambitious Science Teaching, how we engaged preservice teachers in learning to teach, and what we believed we could accomplish as a more explicit community of colleagues. The emergence of this community is not an accident—it is the result of both organically emerging connections from shared interests and intentional efforts to understand each other's thinking, principles, practices, and pedagogies. We accomplish the latter by working together while simultaneously reaching out to new colleagues to develop and extend our networks. Communities require work, but they provide support and create entities that are synergistically more powerful than the individuals who compose them.

We believe our community of teacher education scholars and the work of this volume can serve as a potential model for how teacher educators can develop an inquiry stance on their own practice, and might create a community of codesigners of teacher education pedagogies allowing for research across sites. Self-study has a long tradition in teacher education scholarship, and we have an opportunity of building on that tradition by embedding multiple self-studies in diverse contexts, framed by a shared and articulated vision of teaching and learning. Examining teacher education across contexts is a complex problem, but one that teacher educators must grapple with as there is increasing pressure to justify the value of rigorous teacher preparation programs. Being able to speak to the value of our programs individually is no longer enough to provide evidence that teacher education is necessary in the development of future teachers.

We need to better understand which designed features of our learning environments impact teacher learning, and in what ways such designs are successful and should be improved. Such questions can best be answered by looking across contexts of teacher education. Just as with core practices of science teaching, the enactment of core pedagogies of science teacher education will and should vary across contexts. After all, teacher education is a relational activity just as much as teaching. We see this diversity of enactment as a richness that can help us to better understand the complexity of teacher learning. By having a clearly articulated set of shared practices and pedagogies and collecting data that includes parallel instruments and tools across contexts, we have the opportunity not only to provide evidence of value for teacher education, but also to improve our pedagogies, and by extension the teaching and learning of science in thousands of classrooms across the country. We have reach, but we need stronger footing.

The joy and challenge of teacher education is that our efforts have the possibility of being magnified by orders of magnitude as the teachers we work and learn with are also each working and learning with hundreds, and eventually thousands, of students over their careers. This ripple effect provides our community with a tremendous opportunity to impact not only students' learning of science, but students' understanding of themselves and the relationship they and their community have to the creation of scientific knowledge. It is this opportunity that fuels us, both individually and as a community.

Notes

Introduction

1. National Research Council, *A Framework for K–12 Science Education: Practices, Crosscutting Concepts, and Core Ideas* (Washington, DC: National Academies Press, 2012); NGSS Lead States, *Next Generation Science Standards: For States, by States* (Washington, DC: National Academies Press, 2013).
2. Magdalene Lampert et al., "Using Designed Instructional Activities to Enable Novices to Manage Ambitious Mathematics Teaching," in *Instructional Explanations in the Disciplines,* ed. Mary K. Stein and Linda Kucan (New York: Springer, 2010), 129–41.
3. Kenneth Zeichner, "The Turn Once Again Toward Practice-Based Teacher Education," *Journal of Teacher Education* 63, no. 5 (2012): 376–82.
4. Mark Windschitl and Angela Calabrese Barton, "Rigor and Equity by Design: Seeking a Core of Practices for the Science Education Community," in *AERA Handbook of Research on Teaching*, 5th ed. (Washington, DC: American Educational Research Association, 2016), 1100.
5. Mary Kennedy, "The Role of Pre-service Teacher Education," in *Teaching as the Learning Profession*, ed. Linda Darling-Hammond and Gary Sykes (San Francisco: Jossey-Bass, 1999), 54–85.
6. Pam Grossman and Morva McDonald, "Back to the Future: Directions for Research in Teaching and Teacher Education," *American Educational Research Journal* 45, no. 1 (2008): 184–205.
7. Pam Grossman et al., "Teaching Practice: A Cross-Professional Perspective," *Teachers College Record* 111, no. 9 (2009): 2055–100.
8. Dan Lortie, *School Teacher: A Sociological Inquiry* (Chicago: University of Chicago Press, 1975).

Chapter 1

1. Karen Hammerness, *Seeing Through Teachers' Eyes: Professional Ideals and Classroom Practices* (New York: Teachers College Press, 2006).
2. Maxine Greene, *Releasing the Imagination: Essays on Education, the Arts, and Social Change.* (San Francisco: Jossey-Bass, 2000), 197–98.
3. Pamela Grossman et al., "Teaching Practice: A Cross-Professional Perspective," *Teachers College Record* 11, no. 9 (2009): 2055–100.
4. Morva Macdonald, "The Integration of Social Justice in Teacher Education: Dimensions of Student Teachers' Opportunities to Learn," *Journal of Teacher Education* 56, no. 5 (2005): doi:10.1177/0022487105279569.
5. Linda Darling-Hammond et al., *Preparing Teachers for a Changing World: What Teachers Should Learn and Be Able to Do* (San Francisco: Jossey-Bass, 2005).
6. Mary Kennedy, "The Role of Preservice Teacher Education," in *Teaching as the Learning Profession: Handbook of Policy and Practice*, ed. Gary Sykes and Linda Darling-Hammond (San Francisco: Jossey-Bass, 1999), 54.

7. Magdalene Lampert, "Learning Teaching in, from, and for Practice: What Do We Mean?," *Journal of Teacher Education* 61, no. 1–2 (2010): 21–34, doi:10.1177/002247109347321; Francesca Forzani, "Understanding 'Core Practices' and 'Practice-Based Teacher Education': Learning from the Past," *Journal of Teacher Education* 65, no. 4 (2014), 357–68, doi:10.1177/0022487114533800.

8. Lampert, "Learning Teaching in," 21; Forzani, "Undertanding 'Core Practices,'" 357.

9. Jean Lave and Etienne Wenger, *Situated Learning: Legitimate Peripheral Participation* (Cambridge, UK: Cambridge University Press, 1991).

10. Lampert, "Learning Teaching in," 29.

11. Thomas Philip, "Principled Improvisation to Support Novice Teacher Learning," *Teachers College Record* 121, no. 4 (2019): 4; see also Keith Sawyer, "Creative Teaching: Collaborative Discussion as Disciplined Improvisation," *Educational Researcher* 33, no. 2 (2004).

12. Mike Rose, *The Mind at Work: Valuing the Intelligence of the American Worker* (New York: Penguin Books, 2005).

13. Kurt Lewin, *Field Theory in Social Science: Selected Theoretical Papers* (New York: Harper & Row, 1951).

14. Karen Hammerness et al., "How Teachers Learn and Develop," in *Preparing Teachers for a Changing World: What Teachers Should Learn and Be Able to Do*, ed. Linda Darling-Hammond et al. (San Francisco: Jossey-Bass, 2005), 358.

15. Daniel C. Lortie, *Schoolteacher: A Sociological Study* (Chicago: University of Chicago Press, 1975).

16. Linda Darling-Hammond, *The Flat World and Education: How America's Commitment to Equity Will Determine Our Future* (San Francisco: Jossey-Bass, 2010); Linda Darling-Hammond et al., *Empowered Educators: How High-Performing Systems Shape Teaching Quality Around the World* (San Francisco: Jossey-Bass, 2017).

17. Pam Grossman, "Framework for Teaching Practice: A Brief History of an Idea," *Teachers College Record* 113, no. 12 (2011): 2836.

18. Scott McDonald, "Building a Conversation: Preservice Teachers' Use of Video as Data for Making Evidence-Based Arguments About Practice," *Educational Technology* 1 (2010): 28–31.

19. Marilyn Cochran-Smith and Susan Lytle, *Inquiry as Stance: Practitioner Research for the Next Generation* (New York: Teachers College Press, 2009).

20. Hammerness, *Seeing Through Teachers' Eyes*.

21. Amber Strong Makaiau with Linda Summers Strong, "From School-Culture-to-Family-Culture: Reflections on Four Generations of a Deweyan Education in Hawai'i," *Educational Perspectives* 47, no. 1–2 (2015): 44–49.

22. Dorinda Carter Andrews, "The Hardest Thing to Turn From: The Effects of Service Learning on Preparing Urban Educators," *Equity and Excellence in Education* 42, no. 3 (2009), 272–93; Django Paris and H. Samy Alim, "What Are We Seeking to Sustain Through Culturally Sustaining Pedagogy? A Loving Critique Forward," *Harvard Educational Review* 84, no. 1 (2014): 85–100.

23. Robin Riedy, Gina Tesoriero, and William R. Penuel, "Understanding the Vision for Science in NGSS Adopting and Non-adopting States" (paper presented at the annual meeting for the American Educational Research Association, Toronto, CA, 2018).

24. Sharon Feiman-Nemser and Margret Buchman, "Pitfalls of Experience in Teacher Preparation," *Teachers College Record* 87, no. 1 (1985): 53–65.

25. Marilyn Cochran-Smith, Lexie Grudnoff, and Kari Smith, "Educating Teacher Educators: International Perspectives," *The New Educator* 16, no. 1 (2019): 5–24.

26. Pam Grossman, "Framework for Teaching Practice: A Brief History of an Idea," *Teachers College Record* 113, no. 12 (2011): 2836.

27. Philip, "Principled Improvisation," 1–32.

28. Pam Grossman, ed., *Teaching Core Practices in Teacher Education* (Cambridge: Harvard Education Press, 2018).

29. Morva McDonald, Sarah Kavanaugh, and Elham Kazemi, "Core Practices and Pedagogies of Teacher Education: A Call for a Common Language and Collective Activity," *Journal of Teacher Education* 64, no. 5 (2013), doi:10.1177/0022487113493807.

30. Magdalene Lampert et al., "Keeping It Complex: Using Rehearsals to Support Novice Teacher Learning of Ambitious Teaching," *Journal of Teacher Education* 64, no. 3 (2013): 226–43, doi:10.1177/0022487112473837.

Chapter 2

1. John Hattie, "Teachers Make a Difference: What Is the Research Evidence?" (paper presented at the ACER Research Conference, "Building Teacher Quality: What Does the Research Tell Us," Melbourne, Australia, 2003), http://research.acer.edu.au/research_conference_2003/4; Raj Chetty, John N. Friedman, and Jonah E. Rockoff, "Measuring the Impacts of Teachers II: Teacher Value-Added and Student Outcomes in Adulthood," *American Economic Review* 104, no. 9 (2014): 2633–79.

2. Richard Sohmer, Sarah Michaels, and M. C. O'Connor, "Guided Construction of Knowledge in the Classroom: The Troika of Talk, Tasks and Tools," ed. Baruch Schwarz et al., *Transformation of Knowledge Through Classroom Interaction* (London and New York: Routledge, 2009), 11–137.

3. James V. Hoffman et al., "What Can We Learn from Studying the Coaching Interactions Between Cooperating Teachers and Preservice Teachers? A Literature Review," *Teaching and Teacher Education* 52 (2015): 99–112; Hosun Kang and Mark Windschitl, "How Does Practice-Based Teacher Preparation Influence Novices' First-Year Instruction?," *Teachers College Record* 120, no. 8 (2018): 1–44.

4. Darcy Leigh and Richard Freeman, "Teaching Politics After the Practice Turn," *Politics* 39, no. 3 (2019): 379–92.

5. Keith Sawyer, "What Makes Good Teachers Great? The Artful Balance of Structure and Improvisation," in *Structure and Improvisation in Creative Teaching*, ed. R. K. Sawyer (Cambridge, UK: Cambridge University Press, 2011): 1–24, doi:10.1017/CBO9780511997105.002.

6. Rick A. Duschl, Heidi Schweingruber, and Andrew W. Shouse, eds., *Taking Science to School: Learning and Teaching Science in Grades K–8* (Washington, DC: National Academies Press, 2007).

7. Kathleen Roth and Helen Garnier, "What Science Teaching Looks Like: An International Perspective," *Educational Leadership* 64, no. 4 (2007): 16–23; Iris R. Weiss et al., *Looking Inside the Classroom* (Chapel Hill, NC: Horizon Research, 2003), horizon-research.com/horizonresearchwp/wp-content/uploads/2013/04/complete-1.pdf.

8. M. Pilar Jiménez-Aleixandre et al., "'Doing the Lesson' or 'Doing Science': Argument in High School Genetics," *Science Education* 84, no. 6 (2000): 757–92.

9. Hoffman et al., "What Can We Learn."

10. Mark Windschitl, Jessica Thompson, and Melissa Braaten, *Ambitious Science Teaching* (Boston, MA: Harvard Education Press, 2018).

11. Hosun Kang and Mark Windschitl, "How Does Practice-Based Teacher Preparation Influence Novices' First-Year Instruction?," *Teachers College Record* 120, no. 8 (2018): 1–44.

12. Jessica Thompson, Mark Windschitl, and Melissa Braaten, "Developing a Theory of Ambitious Early-Career Teacher Practice," *American Educational Research Journal* 50, no. 3 (2013): 574–615.

13. Hosun Kang and Doron Zinger, "What Do Core Practices Offer in Preparing Novice Science Teachers for Equitable Instruction?," *Science Education* 103, no. 4 (2019): 823–53.

Chapter 3

1. Django Paris and H. Samy Alim, "What Are We Seeking to Sustain Through Culturally Sustaining Pedagogy? A Loving Critique Forward," *Harvard Educational Review* 84, no. 1 (2014): 85–100.

2. Rochelle Gutierrez, "Enabling the Practice of Mathematics Teachers in Context: Towards a New Equity Research Agenda," *Mathematical Thinking and Learning* 4 no. 2–3 (2002): 145–87.

3. Megan Bang et al., "Desettling Expectations in Science Education," *Human Development* 55 no. 5–6 (2012): 302–18; Beth Warren and Ann S. Rosebery, "Navigating Interculturality: African American Male Students and the Science Classroom," *Journal of African American Males in Education* 2, no. 1 (2011); Na'ilah Nasir, *Racialized Identities: Race and Achievement Among African American Youth* (Redwood City, CA: Stanford University Press, 2011).

4. Cheryl K. Lupenui et al., "Nā Hopena Aʻo HĀ Statements: BREATH," http://www.hawaii publicschools.org/DOE%20Forms/NaHopenaAoE3.pdf, 1.

5. David A. Gruenewald, "Foundations of Place: A Multidisciplinary Framework for Place-Conscious Education," *American Educational Research Journal* 40, no. 3 (2003): 619–54, 625.

6. Patrick Wolfe, "Settler Colonialism and the Elimination of the Native," *Journal of Genocide Research* 8, no. 4 (2006): 387–409.

7. Mary Kawena Pukui, *ʻŌlelo Noʻeau: Hawaiian Proverbs and Poetical Sayings* (Honolulu, HI: Bishop Museum Press Special Publication No. 71, 1983).

8. Pukui, *ʻŌlelo Noʻeau*.

9. Robin DiAngelo, *White Fragility: Why It's So Hard for White People to Talk About Racism* (Boston, MA: Beacon Press, 2018); Lisa Delpit, "The Silenced Dialogue: Power and Pedagogy in Educating Other People's Children," *Harvard Educational Review* 58, no. 3 (1988): 280–99; Bang et al., "Desettling Expectations," 302–18.

10. Victoria B. Costa, "When Science Is 'Another World': Relationships Between Worlds of Family, Friends, School, and Science," *Science Education* 79, no. 3 (1995): 313–33.

11. Julie Brown et al., "Advancing Culturally Responsive Science Education in Secondary Classrooms Through an Induction Course," *International Journal of Designs for Learning* 9, no. 1 (2018): 14–33.

12. Dorinda J. Carter Andrews et al., "Beyond Damage-Centered Teacher Education: Humanizing Pedagogy for Teacher Educators and Preservice Teachers," *Teachers College Record* 121, no. 6 (2019): 1–28; Ana Maria Villegas and Tamara Lucas. "The Culturally Responsive Teacher," *Educational Leadership* 64, no. 6, (2007): 28–33.

13. Ann S. Rosebery, Beth Warren, and Eli Tucker-Raymond, "Developing Interpretive Power in Science Teaching," *Journal of Research in Science Teaching* 53, no. 10 (2016): 1572.

14. Rosebery, Warren, and Tucker-Raymond, "Developing Interpretive Power in Science Teaching."

15. E. Suárez, "Supporting and responding to emerging bilingual students' translanguaging: When reasoning and communicating about natural phenomena." (paper presented at the National Association for Research in Science Teaching, Atlanta, Georgia, March, 2018).

16. Teaching Tolerance, "Social Justice Standards: The Teaching Tolerance Anti-bias Framework," 2016, www.tolerance.org.

17. Alexis D. Patterson, "Equity in Groupwork: The Social Process of Creating Justice in a Science Classroom," *Cultural Studies of Science Education* (2019): 1–21.

Chapter 4

1. Daniel Lortie, *Schoolteacher: A Sociological Study* (Chicago: University of Chicago Press, 1975).

2. Lortie, 79.

3. R. Keith Sawyer, "Creative Teaching: Collaborative Discussion as Disciplined Improvisation," *Educational Researcher* 33, no. 2 (2004): 12–20.

4. Miriam G. Sherin and Rosemary S. Russ, "Making Sense of Teacher Noticing Via Video," in *Digital Video for Teacher Education: Research and Practice*, ed. Brendan Calandra and Peter J. Rich (New York: Routledge, 2014), 3–20.

5. Sherin and Russ.

6. Sharon Feiman-Nemser and Kathrene Beasley, "Mentoring as Assisted Performance: A Case of Co-Planning," in *Constructivist Teacher Education: Building New Understandings*, ed. Virginia Richardson (London: Falmer Press, 1997), 108–26.

7. Pam Grossman et al., "Teaching Practice: A Cross-Professional Perspective," *Teachers College Record* 111, no. 9 (2009): 2055–100.

8. Elizabeth A. Davis et al., "Teaching the Practice of Leading Sense-Making Discussions in Science: Science Teacher Educators Using Rehearsals," *Journal of Science Teacher Education* 28, no. 3 (2017): 275–93; Elham Kazemi et al., "Getting Inside Rehearsals: Insights from Teacher Educators to Support Work on Complex Practice," *Journal of Teacher Education* 67, no. 1 (2016): 18–31.

Chapter 5

1. Daniel C. Lortie, *Schoolteacher: A Sociological Study* (Chicago: University of Chicago Press, 1975).

2. Pamela Grossman and Morva McDonald, "Back to the Future: Directions for Research in Teaching and Teacher Education," *American Educational Research Journal* 45, no. 1 (2008): 184–205.

3. Pamela Grossman et al., "Teaching Practice: A Cross-Professional Study," *Teachers College Record* 111, no. 9 (2009): 2055–100.

4. Eve Manz and Enrique Suarez,"Supporting Teachers to Negotiate Uncertainty for Science, Students, and Teaching," *Science Education* 102, no. 4 (2018): 771–95.

5. David Stroupe and Amelia Wenk Gotwals, "'It's 1000 Degrees in Here When I Teach': Providing Preservice Teachers with an Extended Opportunity to Approximate Ambitious Instruction," *Journal of Teacher Education* 69, no. 3 (2018): 294–306.

6. National Research Council, *A Framework for K–12 Science Education: Practices, Crosscutting Concepts, and Core Ideas* (Washington, DC: National Academies Press, 2012).

7. Hala Ghousseini, "Core Practices and Problems of Practice in Learning to Lead Classroom Discussions," *Elementary School Journal* 115, no. 3 (2015): 334–57.

8. Ilana S. Horn and Sara S. Campbell, "Developing Pedagogical Judgment in Novice Teachers: Mediated Field Experience as a Pedagogy for Teacher Education," *Pedagogies: An International Journal* 10, no. 2 (2015): 149–76; Jennifer Sun and Elizabeth A. van Es, "An Exploratory Study of the Influence That Analyzing Teaching Has on Preservice Teachers' Classroom Practice," *Journal of Teacher Education* 66, no. 3 (2015): 201–14.

9. Amelia Wenk Gotwals and Daniel Birmingham, "Eliciting, Identifying, Interpreting and Responding to Students' Ideas: Teacher Candidates' Growth in Formative Assessment Practices," *Research in Science Education* 46 (2015): 365–88; David Stroupe, "Beginning Teachers' Use of Resources to Enact and Learn from Ambitious Instruction," *Cognition and Instruction* 34, no. 1 (2016): 51–77; Miriam Gamoran Sherin, Victoria R. Jacobs, and Randolph A. Philipp, eds., *Mathematics Teacher Noticing: Seeing Through Teachers' Eyes* (New York: Routledge, 2011).

10. Ilana L. Horn and Britnie D. Kane, "What We Mean When We Talk About Teaching: The Limits of Professional Language and Possibilities for Professionalizing Discourse in Teachers; Conversations," *Teachers College Record* 121, no. 4 (2019).

11. Paul Cobb, "Discussion Remarks" (annual meeting of American Educational Research, Vancouver, Canada, 2011).

12. David Stroupe and Mark Windschitl, "Supporting Ambitious Instruction by Beginning Teachers with Specialized Tools and Practices," in *Newly Hired Teachers of Science: A Better Beginning*, ed. J. Luft and S. Dubois (Netherlands: Sense Publishers, 2015), 181.

13. Paul Black and Dylan Wiliam, "Assessment and Classroom Learning," *Assessment in Education* 5, no. 1 (1998): 7–74.

14. Stephanie Spear, "Watch Award-Winning 'Mountains of the Moon,'" EcoWatch, November 6, 2014, https://www.ecowatch.com/watch-award-winning-mountains-of-the-moon-1881966595.html.

15. Mark Windschitl et al., "Proposing a Core Set of Instructional Practices and Tools for Teachers of Science," *Science Education* 96, no. 5 (2012): 878–903.

16. Thomas M. Philip et al., "Making Justice Peripheral by Constructing Practice as 'Core': How the Increasing Prominence of Core Practices Challenges Teacher Education," *Journal of Teacher Education* 70, no. 3 (2019): 251–64.

Chapter 6

1. Melissa Braaten, "Persistence of the Two-Worlds Pitfall: Learning to Teach Within and Across Settings," *Science Education* 103, no. 1 (2019): 61–91; Magdalene Lampert et al., "Keeping It Complex: Using Rehearsals to Support Novice Teacher Learning of Ambitious Teaching," *Journal of Teacher Education* 64, no. 3 (2013): 226–43.

2. Richard Sohmer et al., "Guided Construction of Knowledge in the Classroom: The Troika of Well-Structured Talk, Tasks, and Tools," in *Advances in Learning and Instruction,* ed. Baruch Schwarz, Tommy Dreyfus, and Rina Hershkowitz (London: Elsevier, 2009), 105–29.

3. Neil Mercer and Karen Littleton, *Dialogue and the Development of Children's Thinking: A Sociocultural Approach* (London: Routledge, 2007).

4. Cynthia Ballenger, "The Puzzling Child: Challenging Assumptions About Participation and Meaning in Talking Science," *Language Arts* 81, no. 4 (2004): 303–11.

5. Sohmer et al., "Guided Construction of Knowledge."

6. Eric R. Banilower et al., *Report of the 2018 NSSME+* (Chapel Hill, NC: Horizon Research, 2018).

7. For examples, see Rita MacDonald, H. Gary Cook, and Emily C. Miller, *Doing and Talking Science: A Teacher's Guide to Meaning-Making with English Learners* (Madison, WI: WCER),

2014, http://stem4els.wceruw.org/resources/WIDA-Doing-and-Talking-Science.pdf; Sarah Michaels and Cathy O'Connor, *Talk Science Primer* (Cambridge, MA: TERC), 2012, http://inquiryproject.terc.edu/shared/pd/TalkScience_Primer.pdf.

8. See Morva McDonald, Elham Kazemi, and Sarah Kavanagh, "Core Practices and Pedagogies of Teacher Education: A Call for a Common Language and Collective Activity," *Journal of Teacher Education* 64, no. 5 (2013): 378–86, doi:10.1177/0022487113493807.

9. Hala Ghousseini, Heather Beasley, and Sarah Lord, "Investigating the Potential of Guided Practice with an Enactment Tool for Supporting Adaptive Performance," *Journal of the Learning Sciences* 24, no. 3 (2015): 461–97; Mark Windschitl, Jessica Thompson, and Melissa Braaten, "Ambitious Pedagogy by Novice Teachers: Who Benefits from Tool-Supported Collaborative Inquiry into Practice and Why?" *Teachers College Record* 113, no. 7 (2011): 1311–60.

10. Lucy Avraamidou and Carla Zembal-Saul, "In Search of Well-Started Beginning Science Teachers: Insights from Two First-Year Elementary Teachers," *Journal of Research in Science Teaching* 47, vol. 6 (2010): 661–86.

11. Braaten, "Persistence of the Two-Worlds Pitfall"; Sharon Feiman-Nemser and Margret Buchman, "Pitfalls of Experience in Teacher Preparation," *Teachers College Record* 87, vol. 1 (1985): 53–65.

12. Ofelia García, Susana Ibarra Johnson, and Kate Seltzer, *The Translanguaging Classroom: Leveraging Student Bilingualism for Learning* (Philadelphia, PA: Caslon, 2017).

Chapter 7

1. Chimamanda Ngozi Adichie, "The Danger of a Single Story," filmed July 2009 at TEDGlobal, video, 18:43, https://www.ted.com/talks/chimamanda_ngozi_adichie_the_danger_of_a_single_story.

2. National Academies of Sciences, Engineering, and Medicine, *Science Teachers' Learning: Enhancing Opportunities, Creating Supportive Contexts* (Washington, DC: National Academies Press, 2015), doi:10.17226/21836.

3. Lucy Avraamidou, "Stories We Live, Identities We Build: How Are Elementary Teachers' Science Identities Shaped by Their Lived Experiences?," *Cultural Studies of Science Education* 14, no. 1 (2018): 1–27, doi:Ó10.1007/s11422-017-9855-8.

4. Lucy Avraamidou, "Studying Science Teacher Identity: An Introduction," in *Studying Science Teacher Identity: Theoretical, Methodological and Empirical Explorations*, ed. Lucy Avraamidou (Rotterdam: SensePublishers, 2016), 1–14, doi:10.1007/978-94-6300-379-7.

5. Kris D. Gutiérrez et al., "Replacing Representation with Imagination: Finding Ingenuity in Everyday Practices," *Review of Research in Education* 41, no. 1 (2017): 30, doi:10.3102/0091732x16687523.

6. Carla Zembal-Saul, "Learning to Teach Elementary School Science as Argument," *Science Education* 93, no. 4 (2009): 687–719, doi:10.1002/sce.20325.

7. Heidi B. Carlone, Julie Haun-Frank, and Sue C. Kimmel, "Tempered Radicals: Elementary Teachers' Narratives of Teaching Science Within and Against Prevailing Meanings of Schooling," *Cultural Studies of Science Education* 5, no. 4 (2010): 941–65, doi:10.1007/s11422-010-9282-6.

8. Felicia Moore Mensah and Iesha Jackson, "Whiteness as a Property in Science Teacher Education," *Teachers College Record* 120, no. 1 (2018): 1–38.

9. Gloria Ladson-Billings, "'Who You Callin' Nappy-Headed?' A Critical Race Theory Look at the Construction of Black Women," *Race Ethnicity and Education* 12, no. 1 (2009): 87–99, doi:10.1080/13613320802651012.

10. Henry A. Giroux and Susan Searls Giroux, "Challenging Neoliberalism's New World Order: The Promise of Critical Pedagogy," *Cultural Studies ↔ Critical Methodologies* 6, no. 1 (2006): 21–32, doi: 10.1177/1532708605282810.

11. Carlone, Haun-Frank, and Kimmel, "Tempered Radicals," 941–65.

12. Kari Kokka, "Radical STEM Teacher Activism: Collaborative Organizing to Sustain Social Justice Pedagogy in STEM Fields," *Educational Foundations* 31, no. 1–2 (2018): 88–114.

13. Mark Windschitl, Jessica Thompson, and Melissa Braaten, *Ambitious Science Teaching* (Cambridge: Harvard Education Press, 2018), 66.

14. April Leuhmann, "Practice-Linked Identity Development in Science Teacher Education: GET REAL! Science as a Figured World," in *Studying Science Teacher Identity*, ed. Lucy Avraamidou (Rotterdam: SensePublishers, 2016), 15–47.

15. Luis C. Moll, Cathy Amanti, Deborah Neff, and Norma Gonzalez, "Funds of Knowledge for Teaching: Using a Qualitative Approach to Connect Homes and Classrooms," *Theory into Practice* 31, no. 2 (1992): 132–41, doi:10.1080/00405849209543534.

16. Megan Hopkins, Carla Zembal-Saul, May Lee, and Jennifer Cody, "Starting Small: Creating a Supportive Context for Professional Learning That Fosters Emergent Bilingual Children's Sensemaking in Elementary Science," in *Supporting Teacher Learning for Sensemaking in Elementary Science*, ed. Elizabeth A. Davis, Carla Zembal-Saul, and Sylvie M. Kademian (New York: Routledge, 2020), 218–32.

17. Eve Manz and Enrique Suárez, "Supporting Teachers to Negotiate Uncertainty for Science, Students, and Teaching," *Science Education* 102, no. 4 (2018): 771–95, doi:10.1002/sce.21343.

18. Mary Manke, review [untitled] of *Classroom Discourse: The Language of Teaching and Learning*, by Courtney Cazden, Language in Society 19, no. 3 (1990): 436–39.

19. Megan Bang et al., "Toward More Equitable Learning in Science," in *Helping Students Make Sense of the World Using Next Generation Science and Engineering Practices*, ed. Christina V. Schwarz, Cynthia Passmore, and Brian J. Reiser (Arlington, VA: NSTA Press, 2017), 33–58.

20. Phillip Bell, "Infrastructuring Teacher Learning About Equitable Science Instruction," *Journal of Science Teacher Education* 30, no. 7 (2019): 681–90, doi:10.1080/1046560X.2019.1668218.

21. Bell.

22. Bryan A. Brown, *Science in the City: Culturally Relevant STEM Education* (Cambridge MA: Harvard Education Press, 2019).

23. Debra E. Meyerson, *Tempered Radicals: How People Use Difference to Inspire Change at Work* (Boston: Harvard Business School Press, 2001), 5.

24. Etienne Wenger, *Communities of Practice: Learning, Meaning, and Identity* (New York: Cambridge University Press, 1999), 110.

25. Wenger, 176.

26. Julie A. Luft and Nancy C. Patterson, "Bridging the Gap: Supporting Beginning Science Teachers," *Journal of Science Teacher Education* 13, no. 4 (2002): 267–82, doi:10.1023/A:1022518815872.

27. Kenneth M. Zeichner, *Teacher Education and the Struggle for Social Justice* (New York: Routledge, 2009).

28. Carla Zembal-Saul et al., "Learning to Teach Science in an Elementary School Professional Development School Partnership," in *Sensemaking in Elementary Science*, ed. Elizabeth A. Davis, Carla Zembal-Saul, and Sylvie M. Kademian (New York: Routledge, 2020), 204–17.

29. Dorothy Holland et al., *Identity and Agency in Cultured Worlds* (Cambridge, MA: Harvard University Press, 1998), 277, 276.

Chapter 8

A portion of this work was supported by funding from the Spencer Foundation, and another portion was funded by a grant from the Bill & Melinda Gates Foundation to TeachingWorks; we are grateful for this funding. Many of the tools and frameworks described in this chapter are available in the TeachingWorks Resource Library at http://library.teachingworks.org as a result of the Gates funding. We have had the opportunity to work with and learn from the many graduate students who have participated in the Elementary Science Methods Planning Group at the University of Michigan over the years. Betsy would also like to recognize how she learned with and from the Elementary Curriculum Design Group in the design of our elementary teacher education program. Most important, we appreciate the experiences we have had with the hundreds of preservice elementary teachers with whom we have had the privilege of working.

1. National Research Council, *A Framework for K–12 Science Education: Practices, Crosscutting Concepts, and Core Ideas* (Washington, DC: National Academies Press, 2012); NGSS Lead States, *Next Generation Science Standards: For States, By States* (Washington, DC: National Academies Press, 2013); Philip Bell, Bruce Lewenstein, Andrew Shouse, and Michael Feder, eds., *Learning Science in Informal Environments: People, Places and Pursuits* (Washington, DC: National Academy Press, 2009).

2. NRC, *Framework*.

3. Deborah Loewenberg Ball and Francesca Forzani, "The Work of Teaching and the Challenge for Teacher Education," *Journal of Teacher Education* 60 (2009): 497–511; Pamela Grossman, ed., *Teaching Core Practices in Teacher Education* (Cambridge, MA: Harvard Education Press, 2018); Pamela Grossman, Karen Hammerness, and Morva McDonald, "Redefining Teaching, Re-imagining Teacher Education," *Teachers and Teaching: Theory and Practice* 15, no. 2 (2009): 273–89.

4. Magdalene Lampert, "Learning Teaching in, from, and for Practice: What Do We Mean?," *Journal of Teacher Education* 61, no. 1–2 (2010): 21–34, doi:10.1177/0022487109347321; Anna Maria Arias and Elizabeth A. Davis, "Supporting Children to Construct Evidence-Based Claims in Science: Individual Learning Trajectories in a Practice-Based Program," *Teaching and Teacher Education* 66 (2017): 204–18.

5. Hala Ghousseini, Heather Beasley, and Sarah Lord, "Investigating the Potential of Guided Practice with an Enactment Tool for Supporting Adaptive Performance," *Journal of the Learning Sciences* 24, no. 3 (2015): 461–97; Grossman, ed., *Teaching Core Practices*; Grossman, Hammerness, and McDonald, "Redefining Teaching"; Mark Windschitl et al., "Proposing a Core Set of Instructional Practices and Tools for Teachers of Science," *Science Education* 96, no. 5 (2012): 878–903.

6. Elizabeth A. Davis and Timothy Boerst, "Designing Elementary Teacher Education to Prepare Well-Started Beginners," *TeachingWorks Working Papers* (Ann Arbor, MI: University of Michigan, 2014), http://www.teachingworks.org/images/files/TeachingWorks_Davis_Boerst_WorkingPapers_March_2014.pdf.

7. Lee S. Shulman, "Those Who Understand: Knowledge Growth in Teaching," *Educational Researcher* 15, no. 2 (1986): 4–14; Deborah Loewenberg Ball, Mark Thames, and Geoffrey Phelps, "Content Knowledge for Teaching: What Makes It Special?," *Journal of Teacher Education* 59, no. 5 (2008): 389–407.

8. Pamela Grossman et al., "Teaching Practice: A Cross-Professional Perspective," *Teachers College Record* 111, no. 9 (2009): 2055–100.

9. Arias and Davis, "Supporting Children"; Elizabeth A. Davis and Annemarie Sullivan Palincsar, "The Development of High-Leverage Science Teaching Practices Among Novice

Elementary Teachers" (paper presented at the annual meeting of the NARST organization, Baltimore, MD, March 31–April 3, 2019).

10. Grossman et al., "Teaching Practice," 2056.

11. Grossman et al., "Teaching Practice," 2069.

12. Davis and Boerst, "Designing Elementary Teacher Education."

13. Matthew Kloser, "Identifying a Core Set of Science Teaching Practices: A Delphi Expert Approach," *Journal of Research in Science Teaching* 51, no. 9 (2014): 1184–217; Windschitl et al., "Proposing a Core Set"; Mark Windschitl and Angela Calabrese Barton, "Rigor and Equity by Design: Locating a Set of Core Teaching Practices for the Science Education Community," in *Handbook of Research on Teaching*, 5th ed., ed. Drew Gitomer and Courtney Bell (Washington, DC: American Educational Research Association, 2016); see also chapter 2, this volume.

14. Roger Bybee et al., *The BSCS 5E Instructional Model: Origins and Effectiveness* (Colorado Springs, CO: Biological Sciences Curriculum Study, 2006).

15. Amanda Benedict-Chambers, "Using Tools to Promote Novice Teacher Noticing of Science Teaching Practices in Post-Rehearsal Discussions," *Teaching and Teacher Education* 59 (2016): 28–44; Sylvie Kademian and Elizabeth A. Davis, "Planning and Enacting Investigation-Based Science Discussions: Designing Tools to Support Teacher Knowledge for Science Teaching," in *Sensemaking in Elementary Science: Supporting Teacher Learning*, ed. Elizabeth A. Davis et al. (New York: Routledge, 2019).

16. The EEE+A framework and other frameworks and tools we have developed for elementary science teacher educators' use are available at the TeachingWorks Resource Library at http://library.teachingworks.org.

17. Davis and Palincsar, "Development of High-Leverage Practices."

18. Arias and Davis, "Supporting Children."

19. Grossman et al., "Teaching Practice," 2055–56.

20. Elizabeth A. Davis, Debra Petish, and Julie Smithey, "Challenges New Science Teachers Face," *Review of Educational Research* 76, no. 4 (2006): 607–51.

21. Eric Banilower et al., *Report of the 2018 NSSME+* (Chapel Hill, NC: Horizon Research, 2018).

22. Ken Appleton, "Science Activities That Work: Perceptions of Primary School Teachers," *Research in Science Education* 32 (2002): 393–410, doi:https://doi.org/10.1023/A:10208 78121184.

23. Carla Zembal-Saul, Katherine McNeill, and Kimber Hershberger, *What's Your Evidence? Engaging K–5 Students in Constructing Explanations in Science* (Boston: Pearson Education, 2013).

24. Kathleen Metz, "Reassessment of Developmental Constraints on Children's Science Instruction," *Review of Educational Research* 65, no. 2 (1995): 93–127.

25. Arias and Davis, "Supporting Children.

26. Grossman et al., "Teaching Practice," 2056.

27. Magdalene Lampert et al., "Keeping it Complex: Using Rehearsals to Support Novice Teacher Learning of Ambitious Teaching," *Journal of Teacher Education* 64, no. 3 (2013): 227.

28. Elham Kazemi, Megan Franke, and Magdalene Lampert, "Developing Pedagogies in Teacher Education to Support Novice Teachers' Ability to Enact Ambitious Instruction," in *Crossing Divides: Proceedings of the 32nd Annual Conference of the Mathematics Education*

Research Group of Australasia, vol. 1, ed. R. Hunter, B. Bicknell, and T. Burgess (2009).

29. Elizabeth A. Davis et al., "Teaching the Practice of Leading Sensemaking Discussions in Science: Science Teacher Educators Using Rehearsals," *Journal of Science Teacher Education* 28, no. 3 (2017): 275–93, doi:10.1080/1046560X.2017.1302729.

30. Elizabeth A. Davis, "Approximations of Practice: Scaffolding for Novice Teachers," in *Sensemaking in Elementary Science: Supporting Teacher Learning*, ed. Elizabeth A. Davis et al. (New York: Routledge, 2019).

31. Grossman et al., "Teaching Practice," 2083.

32. Lampert et al., "Keeping It Complex."

33. Benedict-Chambers, "Using Tools."

34. Davis and Palincsar, "Development of High-Leverage Practices."

35. Arias and Davis, "Supporting Children."

36. Lampert, "Learning Teaching"; Arias and Davis, "Supporting Children."

37. Lampert, "Learning Teaching."

38. Ball and Forzani, "The Work of Teaching"; Davis and Boerst, "Designing Elementary Teacher Education."

39. Arias and Davis, "Supporting Children."

40. Lampert, "Learning Teaching."

41. Davis et al., "Teaching the Practice of Leading Sensemaking Discussions"; Lampert et al., "Keeping It Complex."

42. Lampert, "Learning Teaching."

43. Davis and Boerst, "Designing Elementary Teacher Education."

44. NRC, *Framework.*

45. Banilower et al., "NSSME+."

46. Arias and Davis, "Supporting Children"; Lampert, "Learning Teaching."

Chapter 9

1. C. Aaron Price and A. Chiu, "An Experimental Study of a Museum-Based, Science PD Programme's Impact on Teachers and Their Students," *International Journal of Science Education* 40, no. 9 (2018): 941–60.

2. Samuel J. M. M. Alberti, "The Status of Museums: Authority, Identity, and Material Culture," in *Geographies of Nineteenth-Century Science*, ed. David N. Livingstone and Charles W. J. Withers (Chicago: University of Chicago Press, 2011), 51–72.

3. Emily Dawson, "'Not Designed for Us': How Science Museums and Science Centers Socially Exclude Low-Income, Minority Ethnic Groups," *Science Education* 98, no. 6 (2014): 981–1008.

4. Emily Dawson, "Reimagining Publics and (Non) Participation: Exploring Exclusion from Science Communication Through the Experiences of Low-Income, Minority Ethnic Groups," *Public Understanding of Science* 27, no. 7 (2018): 772–86; Noah Weeth Feinstein and David Meshoulam, "Science for What Public? Addressing Equity in American Science Museums and Science Centers," *Journal of Research in Science Teaching* 51, no. 3 (2014): 368–94.

5. Pamela Grossman et al., "Teaching Practice: A Cross-Professional Perspective," *Teachers College Record* 111, no. 9 (2009): 2055–100.

6. Jennifer Adams and Preeti Gupta, "Informal Science Institutions and Learning to Teach: An Examination of Identity, Agency, and Affordances," *Journal of Research in Science Teaching* 54, no. 1 (2017): 121–38.

7. Morva McDonald, Elham Kazemi, and Sara Kavanagh, "Core Practices and Pedagogies of Teacher Education: A Call for a Common Language and Collective Activity," *Journal of Teacher Education* 20, no. 10 (2013): 1–9.
8. National Research Council, *Ready, Set, SCIENCE!: Putting Research to Work in K–8 Science Classrooms*, ed. Sarah Michaels, Andrew W. Shouse, and Heidi A. Schweingruber (Washington, DC: National Academies Press, 2008), doi:10.17226/1182; Sarah Michaels and Cathy O'Connor, *Talk Science Primer* (Cambridge, MA: TERC, 2012).
9. Richard White and Richard Gunstone, "Prediction-Observation-Explanation," in *Probing Understanding* (London: Routledge, 1992): 44–65.
10. Jessica Thompson, "A/B Partner Talk Protocol," *Tools for Ambitious Teaching*, November 12, 2016, https://ambitiousscienceteaching.org/ab-partner-talk-protocol/.
11. Harvard Graduate School of Education, Project Zero, 2009.

Chapter 10

1. John Seely Brown, Allan Collins, and Paul Duguid, "Situated Cognition and the Culture of Learning," *Educational Researcher* 18, no. 1 (1989): 32–42.
2. Dan C. Lortie, *Schoolteacher: A Sociological Study* (Chicago, IL: University of Chicago Press, 1975).
3. Pamela L. Grossman and Morva McDonald, "Back to the Future: Directions for Research in Teaching and Teacher Education," *American Educational Research Journal* 45, no. 1 (2008): 184–205.
4. Gordon Macleod, "Microteaching: End of a Research Era?" *International Journal of Educational Research* 11, no. 5 (1987): 531–41; Pamela Grossman, "The Teaching of Practice in Teacher Education," in *Making a Difference: Challenges for Teachers, Teaching and Teacher Education*, ed. Jude Butcher and Lorraine McDonald (Leiden, Netherlands: Brill | Sense, 2007), 55–65.
5. Magdelene Lampert et al., "Keeping it Complex: Using Rehearsals to Support Novice Teacher Learning of Ambitious Teaching," *Journal of Teacher Education* 64, no. 3 (2013): 226–43.
6. Sharon Feiman-Nemser and Margret Buchmann, "When Is Student Teaching Teacher Education?," *Teaching and Teacher Education* 3, no 4 (1987): 255–73.
7. Koeno Gravemeijer and Paul Cobb, "Design Research from a Learning Design Perspective," in *Educational Design Research*, ed. Jan van den Akker et al. (London and New York: Routledge, 2006), 17–51.

Chapter 11

1. Linda Darling-Hammond, "Teacher Education and the American Future" (Charles W. Hunt Lecture, annual meeting of the American Association of Colleges for Teacher Education, Chicago, Illinois, 2009); Ken Zeichner, "Rethinking the Connections Between Campus Courses and Field Experiences in College- and University-Based Teacher Education," *Journal of Teacher Education* 61, no. 1–2 (2010): 89–99, doi:10.1177/0022487 109347671.
2. Larry Cuban, *Inside the Black Box of Classroom Practice: Change Without Reform in American Education* (Cambridge, MA: Harvard Education Press, 2013); Melissa Braaten, "Persistence of the Two-Worlds Pitfall: Learning to Teach Within and Across Settings," *Science Education* 103, no. 1 (2019): 61–91; Sharon Feiman-Nemser and Buchmann Margret, "Pitfalls of Experience in Teacher Preparation," *Teachers College Record* 87, no. 1 (1985): 53.

3. Randi N. Stanulis et al., "Mentoring as More Than 'Cheerleading': Looking at Educative Mentoring Practices Through Mentors' Eyes," *Journal of Teacher Education* 70, no. 5 (2018): 1–14, doi:10.1177/0022487118773996.

4. Sharon Feiman-Nemser, "From Preparation to Practice: Designing a Continuum to Strengthen and Sustain Teaching," *Teachers College Record* 103, no. 6 (2001): 1013–55.

5. Braaten, "Persistence," 78.

6. Jessica Thompson et al., "Problems Without Ceilings: How Mentors and Novices Frame and Work on Problems-of-Practice," *Journal of Teacher Education* 66, no. 4 (2015): 363–81.

7. Megan Bang and Douglas Medin, "Cultural Processes in Science Education: Supporting the Navigation of Multiple Epistemologies," *Science Education* 94, no. 6 (2010): 1008–26, doi:10.1002/sce.20392; Charlotte L. Land, "Examples of c/Critical Coaching: An Analysis of Conversation Between Cooperating and Preservice Teachers," *Journal of Teacher Education* 69, no. 5 (2018): 493–507, doi:10.1177/0022487118761347.

8. Land, 504.

9. Thompson et al., "Problems Without Ceilings," 378.

10. Cuban, *Inside the Black Box of Classroom Practice.*

11. April L. Luehmann, "Practice-Linked Identity Development in Science Education: Get Real! Science as a Figured World," in *Studying Science Teacher Identity: Theoretical, Methodological and Empirical Exploration,* ed. Lucy Avraamidou (Rotterdam, Netherlands: Sense Publishers, 2016), 15–47.

12. Joshua L. Glazer and Donald J. Peurach, "Occupational Control in Education: The Logic and Leverage of Epistemic Communities," *Harvard Educational Review* 85, no. 2 (2015): 172–202.

13. James P. Gee. "What Video Games Have to Teach Us About Learning and Literacy," *Computers in Entertainment* 1, no. 1 (2003): 2; April L. Luehmann, "Identity Development as a Lens to Science Teacher Preparation," *Science Education* 91, no. 5 (2007): 822–39, doi:10.1002/sce.20209.

14. Django Paris and H. Samy Alim, "What Are We Seeking to Sustain Through Culturally Sustaining Pedagogy? A Loving Critique Forward," *Harvard Educational Review* 84, no. 1 (2014): 85–100.

15. Cynthia E Coburn, William R. Penuel, and Kimberly E. Geil, *Research-Practice Partnerships: A Strategy for Leveraging Research for Educational Improvement in School Districts* (New York: William T. Grant Foundation, 2013), https://wtgrantfoundation.org/library/uploads/2015/10/Research-Practice-Partnerships-at-the-District-Level.pdf; Braaten, "Persistence," 61.

Chapter 12

1. Lev Vygotsky, "Thinking and Speech," in *The Collected Works of L. S. Vygotsky, Volume 1: Problems of General Psychology,* ed. Robert W. Carton and Aaron S. Rieber (New York: Plenum, 1987), 37–285; James Wertsch, *Mind as Action* (Oxford: Oxford University Press, 1998); Sanne F. Akkerman and Arthur Bakker, "Boundary Crossing and Boundary Objects," *Review of Educational Research* 81, no. 2 (2011): 132–69, doi:10.3102/0034654311404435.

2. Michael Cole and Yrjo Engeström, "A Cultural-Historical Approach to Distributed Cognition," in *Distributed Cognitions: Psychological and Educational Considerations,* ed. Gavriel Salomon (Cambridge, UK: Cambridge University Press, 2003), 10.

3. Emily van Zee, Mary Long, and Mark Windschitl, "Secondary Science Teaching Methods Courses," in *The Continuum of Secondary Science Teacher Preparation* (Lieden, Netherlands: Brill|Sense, 2009), 33–47, doi: 10.1163/9789087908041_004.

4. David Stroupe and Amelia Wenk Gotwals, "'It's 1000 Degrees in Here When I Teach': Providing Preservice Teachers with an Extended Opportunity to Approximate Ambitious Instruction," *Journal of Teacher Education* 69, no. 3 (2018): 294–306, doi:10.1177/002248 7117709742.

5. National Research Council, *A Framework for K–12 Science Education: Practices, Crosscutting Concepts, and Core Ideas* (Washington, DC: National Academies Press, 2011), doi:10.17226/18290; NGSS Lead States, *Next Generation Science Standards: For States, By States* (Washington, DC: National Academies Press, 2013), doi:10.17226/18290.

6. Sarah Michaels and Catherine O'Connor, *Talk Science Primer* (Cambridge, MA: TERC, 2012); Jennifer Cartier et al., *5 Practices for Orchestrating Productive Science Discussions* (Reston, VA: National Council of Teachers of Mathematics and Corwin Press, 2013); National Research Council, *Ready, Set, SCIENCE!: Putting Research to Work in K–8 Science Classrooms*, ed. Sarah Michaels, Andrew W. Shouse, and Heidi A. Schweingruber (Washington, DC: National Academies Press, 2008), doi:10.17226/11882.

7. Grant P. Wiggins and Jay McTighe, *Understanding by Design Framework* (Alexandria, VA: Association for Supervision and Curriculum Development, 2012); Cartier et al., *5 Practices*.

8. M. Suzanne Donovan and John D. Bransford, *How Students Learn: Mathematics in the Classroom* (Washington, DC: National Academies Press, 2005), doi:10.17226/11101; National Research Council, *Ready, Set, SCIENCE!*; <<AU: Is this the NRC cite meant? It's 2008 above.>>Todd Campbell, Christina V. Schwarz, and Mark Windschitl, "What We Call Misconceptions May Be Necessary Stepping-Stones Toward Making Sense of the World," *Science and Children* 53, no. 7 (2016): 69–74.

9. Erin Furtak, *Formative Assessment for Secondary Science Teachers* (Thousand Oaks, CA: Corwin Press, 2009).

10. Alberto J. Rodriguez, "What About a Dimension of Engagement, Equity, and Diversity Practices? A Critique of the Next Generation Science Standards," *Journal of Research in Science Teaching* 52, no. 7 (2015): 1031–51, doi:10.1002/tea.21232; Beth Warren et al., "Rethinking Diversity in Learning Science: The Logic of Everyday Sense-Making," *Journal of Research in Science Teaching* 38, no. 5 (May 1, 2001): 529–52, doi:10.1002/tea.1017; Gloria Ladson-Billings, "Culturally Relevant Pedagogy 2.0: A.k.a. the Remix," *Harvard Educational Review* 84, no. 1 (2014): 74–84, doi:10.17763/haer.84.1.p2rj131485484751.

11. Karen Hammerness et al., "How Teachers Learn and Develop," in *Preparing Teachers for a Changing World: What Teachers Should Learn and Be Able to Do*, ed. L. Darling-Hammond and J. Bransford (Hoboken, NJ: Jossey Bass, 2005), 382–89.

12. Sharon Feiman-Nemser and Margret Buchman, "Pitfalls of Experience in Teacher Preparation," *Teachers College Record* 87, no. 1 (1985): 53–65.

13. Dorothea Anagnostopoulos, Emily R. Smith, and Kevin G. Basmadjian, "Bridging the University–School Divide: Horizontal Expertise and the 'Two-Worlds Pitfall,'" *Journal of Teacher Education* 58, no. 2 (March 5, 2007): 138–52, doi:10.1177/0022487106297841.

14. Melissa Braaten and Manali Sheth, "Tensions Teaching Science for Equity: Lessons Learned from the Case of Ms. Dawson," *Science Education* 101, no. 1 (2016): 1–31, doi:10.1002/sce.1254; Mark Windschitl, "Framing Constructivism in Practice as the Negotiation of Dilemmas: An Analysis of the Conceptual, Pedagogical, Cultural, and Political Challenges Facing Teachers," *Review of Educational Research* 72, no. 2 (2002): 131–75, doi:10.3102/00346543072002131.

15. Nathaniel Gage, *The Scientific Basis of the Art of Teaching* (New York: Teachers College Press, 1978).

16. Hammerness et al., "How Teachers Learn."

17. Magdalene Lampert et al., "Keeping It Complex: Using Rehearsals to Support Novice Teacher Learning of Ambitious Teaching," *Journal of Teacher Education* 64, no. 3 (2013): 226–43, doi:10.1177/0022487112473837; Kenneth Zeichner, "The Turn Once Again Toward Practice-Based Teacher Education," *Journal of Teacher Education* 63, no. 5 (2012): 376–82, doi:10.1177/0022487112445789.

18. Miriam Sherin and Sandra Y Han, "Teacher Learning in the Context of a Video Club," *Teaching and Teacher Education* 20, no. 2 (2004): 163–83, doi:10.1016/JTATE.2003.08.001; Miriam Sherin and Elizabeth van Es, "A New Lens on Teaching: Learning to Notice," *Mathematics Teaching in the Middle School* 9, no. 2 (2003): 92–95.

19. Braaten and Sheth, "Tensions Teaching Science."

Chapter 13

1. Cynthia E. Coburn and William R. Penuel, "Research–Practice Partnerships in Education: Outcomes, Dynamics, and Open Questions," *Educational Researcher* 45, no. 1 (2016): 48–54.

2. Anthony S. Bryk et al., *Learning to Improve: How America's Schools Can Get Better at Getting Better* (Cambridge, MA: Harvard Education Press, 2015).

3. Bryk et al.

4. Sharon Feiman-Nemser and Margret Buchmann, *The First Year of Teacher Preparation: Transition to Pedagogical Thinking?* (East Lansing, MI: Institute for Research on Teaching, Michigan State University, 1985).

5. William R. Penuel, "Infrastructuring as a Practice of Design-Based Research for Supporting and Studying Equitable Implementation and Sustainability of Innovations," *Journal of the Learning Sciences* 28, no. 4–5 (2019): 659–77, doi:10.1080/10508406.2018.1552151.

6. Cynthia E. Coburn, "Rethinking Scale: Moving Beyond Numbers to Deep and Lasting Change," *Educational Researcher* 32, no. 6 (2003), doi:10.3102/0013189X032006003; Michael Fullan, *The New Meaning of Educational Change*, 4th ed. (New York: Teachers College Press, 2007).

7. Coburn and Penuel, "Research–Practice Partnerships."

8. Okhee Lee, Helen Quinn, and Guadalupe Valdés, "Science and Language for English Language Learners in Relation to Next Generation Science Standards and with Implications for Common Core State Standards for English Language Arts and Mathematics," *Educational Researcher* 42, no. 4 (2013): 223–33, doi:10.3102/0013189X13480524.

9. Jessica J. Thompson et al., "Toward a Practice-Based Theory for How Professional Learning Communities Engage in the Improvement of Tools and Practices for Scientific Modeling," *Science Education* 103, no. 6 (2019), doi:10.1002/sce.21547.

10. Catherine Lewis, Rebecca Perry, and Aki Murata, "How Should Research Contribute to Instructional Improvement? The Case of Lesson Study," *Educational Researcher* 35, no. 3 (2006): 3–14, doi:10.3102/0013189X035003003.

11. Jana Echevarría, MaryEllen Vogt, and Deborah Short, *Making Content Comprehensible for English Learners: The SIOP® Model*, 4th ed., SIOP model series (Boston: Allyn and Pearson, 2013); Lee, Quinn, and Valdés, "Science and Language for English Language."

12. Jessica Thompson et al., "Launching Networked PLCs: Footholds into Creating and Improving Knowledge of Ambitious and Equitable Teaching Practices in an RPP," *AERA Open,* September 20, 2019, doi:10.1177/2332858419875718.

13. Lora Cohen-Vogel et al., "A Model of Continuous Improvement in High Schools: A Process for Research, Innovation Design, Implementation, and Scale," *Teachers College Record* 118, no. 13 (2016): 1–26.

14. Susan Loucks-Horsley et al., *Designing Professional Development for Teachers of Science and Mathematics*, 3rd ed. (Thousand Oaks, CA: Corwin Press, 2009), 26.

15. See for example, Edward Fergus, *Solving Disproportionality and Achieving Equity: A Leader's Guide to Using Data to Change Hearts and Minds* (Thousand Oaks, CA: Corwin Press, 2017).

16. Penuel, "Infrastructuring as a Practice."

17. Bryk et al., *Learning to Improve.*

18. William R. Penuel et al., "Organizing Research and Development at the Intersection of Learning, Implementation, and Design," *Educational Researcher* 40, no. 7 (2011): 331–37, doi:10.3102/0013189X11421826.

19. James Hiebert and James W. Stigler, "Teaching Versus Teachers as a Lever for Change: Comparing a Japanese and a U.S. Perspective on Improving Instruction," *Educational Researcher* 46, no. 4 (2017): 169–76 doi:10.3102/0013189X17711899; James W. Stigler and James Hiebert, *The Teaching Gap: Best Ideas from the World's Teachers for Improving Education in the Classroom* (New York: Free Press, 1999).

20. For example, see Douglas B. Larkin, Scott C. Seyforth, and Holly J. Lasky, "Implementing and Sustaining Science Curriculum Reform: A Study of Leadership Practices Among Teachers Within a High School Science Department," *Journal of Research in Science Teaching* 4, no. 7 (2009): 813–35, doi:10.1002/tea.20291; Douglas B. Larkin, "Putting Physics First: Three Case Studies of High School Science Department and Course Sequence Reorganization," *School Science & Mathematics* 116, no. 4 (2016): 225–35, doi:10.1111/ssm.12168.

Conclusion

1. Mark Windschitl and David Stroupe, "The Three-Story Challenge: Implications of the Next Generation Science Standards for Teacher Preparation," *Journal of Teacher Education* 68, no. 3 (2017): 251–61.

2. Dan Lortie, *Schoolteacher: A Sociological Inquiry* (Chicago: University of Chicago Press, 1975).

About the Contributors

Naina Abowd, senior manager of curriculum and teaching in the Youth Initiatives department at the American Museum of Natural History (AMNH), manages educators and oversees the development of all the curricula across middle and high school programs at the museum. She oversees professional learning for the educators and teaches the summer residency class for the AMNH MAT program. She is passionate about creating opportunities for underserved students across the city to take advantage of all of the programs and experiences AMNH has to offer. Prior to working at AMNH, Abowd taught high school science at the Urban Assembly School for Green Careers, a public high school in New York City. She received her bachelor's degree from Cornell University as a biological sciences major, with a concentration in neuroscience and behavior, and earned her master's degree in science education from New York University.

Kathryn M. Bateman, PhD, is a postdoctoral researcher with the Research in Spatial Cognition Lab at Temple University. The work in Bateman's chapter was conducted as part of her doctoral studies at the Pennsylvania State University. Her research centers on the nexus between teacher education, educational policy, and geoscience education. Currently, Bateman collaborates with engineers, psychologists, and geologists to examine the use of unmanned aerial vehicles in both geological research and geoscience education, focusing on how geoscientists practice. Her other work focuses on understanding how researchers can collaborate with school communities to navigate policy barriers (real and perceived) to better science teaching and learning in equitable ways through mapping communities, social network analysis, and ethnography.

Melissa Braaten, PhD, is an assistant professor of science education at the University of Colorado Boulder and a former elementary, middle, and high school science teacher. She works closely with teachers and teacher educators to understand how people learn to teach science and to improve science teaching so that all young people experience intellectually engaging and rigorous science education that is humanizing and culturally sustaining. Her research works across multiple contexts for teacher learning, including the preservice preparation of teachers in coursework and field experiences, designed settings such as professional development for teachers, and learning that takes place "on the job" in the day-to-day work of teachers in schools.

Michelle N. Brown is a PhD candidate in science education at Pennsylvania State University. She is interested in understanding how white, monolingual teachers and researchers move toward asset perspectives when considering minoritized students and families. In her current work in a new immigrant, emergent bilingual community, Brown is collaborating with elementary teachers to interrupt deficit perspectives and engage families in meaningful partnerships through science. Her work employs collaborative ethnography and considers how affect, counternarratives, and discursive moves in science learning contexts influence changes in identity and asset perspectives. Prior to this work, Brown was a middle and high school science teacher, STEM coach, and educational equity consultant.

Todd Campbell, PhD, is the department head of curriculum and instruction and a professor of science education in the Neag School of Education at the University of Connecticut. He teaches secondary science methods courses. His research focuses on cultivating imaginative and equitable representations of STEM activity. This is accomplished in formal science learning environments through partnering with preservice and in-service science teachers and leaders to collaboratively focus on supporting student use of modeling as an anchoring epistemic practice to reason about events that happen in the natural world. This work extends into informal learning environments through a focus on iterative design of informal learning spaces and equity-focused STEM identity research.

Heidi Carlone, PhD, is the Hooks Distinguished Professor of STEM Education at the University of North Carolina Greensboro. She works to make science and engineering pathways more accessible and equitable for historically underserved and underrepresented populations by leveraging insights from research and practice. Three questions drive her work: (1) How can innovative K–12 science and engineering instruction cultivate STEM identities for diverse youth? (2) How can we enrich K–12 diverse youths' STEM learning ecologies (support systems) in sustainable ways? (3) How can we design professional learning networks to support and nurture rigorous and equitable STEM teaching? Her studies of equity, identity, and culture in science education have been funded by the National Science Foundation and other nonprofit organizations.

Carolyn Colley, PhD, is currently a science instructional coach at Sartori Elementary in the Renton Public School District in Renton, Washington. In this role, she develops elementary science curriculum and facilitates teachers' professional learning with a constant focus on creating opportunities for elementary students' collective and individual sensemaking in science. Prior to her current position, she earned her doctorate in curriculum and instruction with a specialization in science education from the University of Washington, Seattle. Her research interests include studying conditions that support rigorous and equitable student sensemaking talk and analyzing when professional learning activities, like instructional coaching, help elementary teachers improve science learning opportunities for their students. Her past teaching experiences include fourth-grade bilingual classroom teacher, district science instructional coach, and course instructor for elementary science teaching methods courses at both Antioch University Seattle and the University of Washington.

Elizabeth A. (Betsy) Davis, PhD, is a science educator, teacher educator, and learning scientist at the University of Michigan. She is especially interested in beginning elementary teachers, teachers learning to engage in rigorous and consequential science teaching, and the roles of curriculum materials and practice-based teacher education in promoting teacher learning.

Erin Marie Furtak, PhD, is professor of STEM education and associate dean of faculty at the University of Colorado Boulder. Challenged while trying to realize educational reforms as a public high school biology teacher, Furtak transitioned into a career studying how science teachers learn and improve their daily classroom practices through formative assessment. In a series of multiple studies, Furtak has been partnering with teachers, schools, and districts to learn how teachers can design, enact, and take instructional action on the basis of classroom assessments that they also design. Furtak received the 2011 Presidential Early Career Award for Scientists and Engineers and the German Chancellor Fellowship from the Alexander von Humboldt Foundation (2006). Her research and professional writing has been published in multiple journal articles, research- and practitioner-oriented books, book chapters, humorous essays, and advice columns. She also conducts extensive service to the teaching profession through long-term research and professional development partnerships with school districts and organizations in Colorado and across the United States.

Amelia Wenk Gotwals, PhD, is an associate professor in the Department of Teacher Education at Michigan State University. She is a former middle and high school science teacher. Her current research has two main strands. First, she explores how to support preservice teachers' enactment of Ambitious Science Teaching, specifically as it is related to formative assessment practices. Second, she focuses on working with early elementary teachers as they learn to make space for and support sensemaking science talk in their classrooms.

Ron Gray, PhD, is an associate professor of science education and coordinator of graduate programs in the Center for Science Teaching and Learning at Northern Arizona University. His research interests include beginning and experienced science teacher learning. He is especially interested in Ambitious Science Teaching as an instructional framework and model-based inquiry as a curricular framework for secondary science classrooms. Beyond teacher learning, he is interested in the science studies literature and its potential impact on science education. A former middle school science teacher, Gray received his PhD in science education from Oregon State University.

Preeti Gupta, PhD, is director for youth learning and research at the American Museum of Natural History, where she is responsible for strategic planning and program development for out-of-school youth initiatives. She leads a research agenda centered on youth learning and serves on the faculty for the Masters of Arts in Teaching Earth Science Residency Program. Prior to this she served as senior vice president for education and family programs at the New York Hall of Science. She has a bachelor's degree in bioengineering from Columbia University, a master's degree in education from the George Washington University, and a doctoral degree in urban education from the City University of New York Graduate Center. In 1999, she was selected as one of forty-two women who have contributed to the community in unique ways, showcased in a photo exhibition, the Many Faces of Queens Women. In 2005, she won the inaugural national Roy L. Schafer Leading Edge Award for Experienced Leadership in the Field from the Association for Science Technology Centers. Her research interests include science teacher preparation, youth employment and workforce development, and the role of cultural institutions in mediating identity development in youth.

James Brian Hancock II, PhD, is an instructor of teacher education at Alma College in Michigan. Prior to pursuing his doctorate, Hancock taught high school physics and was a lecturer in the Department of Physics at Central Michigan University. Hancock's research interests include examining how preservice science teachers learn to work on and with student thinking across methods courses and student teaching contexts. He also enjoys supporting—and learning alongside—practicing teachers enacting ambitious science instruction.

Elaine Howes, PhD, is a faculty member in education in the American Natural History Museum's Master of Arts in Teaching Earth Science Residency Program. Her career in education includes four years of high school teaching and umpteen years as a teacher educator and researcher at Teachers College, Columbia University; the University of South Florida; and the American Museum of Natural History's Gilder Graduate School. Immediately before joining the MAT faculty at AMNH, Howes developed curriculum, implemented professional development, and conducted research at BSCS Science Learning in Colorado Springs. Howes's work studying her own teaching, as well as teaching and collaborating with preservice and in-service science teachers, has led to publications about teachers' practices in working with English language learners in science, and the challenges involved in developing environmentally and culturally relevant science curriculum for urban K–12 classrooms. Her current research examines how AMNH MAT graduates learn about their students' ideas, communities, and cultures, and how they use what they learn to inform their earth science teaching. As a member of the AMNH MAT faculty, she is continuing her commitment to working with teachers to develop science education that supports all students in succeeding in science in high-need schools.

Heather Johnson, PhD, is an associate professor of the practice of science education in the Department of Teaching and Learning at Peabody College at Vanderbilt University. Her research explores supports for science teacher learning and how these affect teacher practice and ultimately student learning. These supports include everything from the design of educative curriculum materials to professional development, coaching, and university courses. She is also the principal investigator of two NSF-funded Noyce Teacher Scholarship Programs, both collaborative efforts between Vanderbilt University, Fisk University, and Metropolitan Nashville Public Schools to create a sustainable pipeline for STEM teachers who want to teach in high-need secondary schools. Merging these interests of recruiting and preparing more teachers to teach STEM in high-need schools has brought Johnson to explore ways to help teachers develop and maintain linguistically and culturally sustaining pedagogical practices from teacher preparation and the secondary science classroom.

Rosamond Kinzler, PhD, senior director for science education at the American Museum of Natural History, leads a variety of digital education projects at the museum. For example, she recently served as executive producer of the Hayden Planetarium show Worlds Beyond Earth, which premiered in January 2020 and will be shown in planetariums around the world. In addition, she is the codirector of the museum's Master of Arts in Teaching Earth Science Residency Program, which prepares new earth science teachers. She received her PhD from the Massachusetts Institute of Technology and researched processes by which magmas form and cool, working at Columbia's Lamont Doherty Earth Observatory and in the museum's Earth and Planetary Sciences Department, and is cocurator of the museum's Gottesman Hall of Planet Earth. In her role as codirector of the MAT program, she is interested in the intersection between strong earth science content knowledge and effective teaching practices for diverse students, and how to develop both in residents.

Matthew Kloser, PhD, is the founding director of the University of Notre Dame Center for STEM Education and an associate professor and fellow of the Institute for Educational Initiatives at the University of Notre Dame. Kloser's research focuses on issues of teaching, learning, and assessment in science classrooms with a special focus on biology education. His research includes experimental studies that identify affordances and constraints of learning biology from different text types, mixed methods studies focused on assessment implications for student outcomes, and the relationship between core instructional practices and student outcomes. Kloser received his MEd from the University of Notre Dame and taught high school physics and math for five years before earning his MS in biology and PhD in science education from Stanford University.

Douglas B. Larkin, PhD, is an associate professor in the Department of Teaching and Learning at Montclair State University in New Jersey. He worked as a high school physics and chemistry teacher for ten years—most recently in Trenton, New Jersey—and also served as a Peace Corps volunteer teaching physics and mathematics in Kenya and Papua New Guinea. He received his PhD in teacher education in 2010 from the University of Wisconsin–Madison. His research concerns the preparation of science teachers for culturally diverse classrooms as well as issues of equity and justice in teacher preparation. His most recent book is *Teaching Science in Diverse Classrooms: Real Science for Real Students* (Routledge 2019). He currently serves as an editor for the Science Teacher Education section of the journal *Science Education* and runs the Induction and Mentoring Programs for the Retention of Science Teachers (IMPREST) project at Montclair State University with funding from the National Science Foundation.

April Luehmann, PhD, is an associate professor in teaching and curriculum at the Warner Graduate School of Education, University of Rochester. She designed and directed the secondary science teacher education program, Get Real! Science, which capitalizes on informal learning-to-teach experiences in camps and clubs as precursors and core complements to formal learning-to-teach experiences. The core commitment of the Get Real! Science program is the realization of culturally sustaining pedagogy to support all students (especially those from groups who have been historically underserved) in seeing and realizing the value of learning science for their own lives and purposes. Her research focuses on science teacher development, youth identity development, and the value of augmented reality and other multimodal tools for supporting learners and teachers to author science while taking up uncommon roles in learning together.

Lisa Lundgren, PhD, is a postdoctoral research associate in the department of curriculum and instruction within the Neag School of Education at the University of Connecticut. She teaches courses on conducting educational research for secondary STEM teachers. Her research focuses on the design, development, implementation, and research of social learning within formal, informal, and online environments. She uses a design-based research perspective to develop practical and theoretical insights concerning learning processes within the field of STEM education. Her current projects include using social network analysis to understand connections within levels of the educational system as people implement NGSS and operationalizing the communities of practice theoretical framework for online scientific learning environments.

John-Carlos Marino, PhD, is a supervisor in the Office of Curriculum and Instruction at Detroit Public Schools Community District. He earned his doctorate at the University of Michigan, where he studied elementary students' argumentation in science and history and supported preservice teachers. He currently works on districtwide initiatives supporting classroom instruction and teachers' practice, and is interested in how research-based teaching practices can be applied at scale.

Kavita Kapadia Matsko, PhD, is associate professor and associate dean for teacher education at Northwestern University's School of Education and Social Policy. She has worked in the field of education in the Chicagoland area for over twenty years as a teacher, mentor, teacher educator, and researcher. Her research interests focus on new teacher preparation, mentoring, and induction, with a particular interest in the features of clinical preparation that promote teacher readiness. Matsko received the 2015 Outstanding Article Award from the *Journal of Teacher Education* for work on context-specific teacher preparation, which highlighted the work of the University of Chicago's Urban Teacher Education Program, where she served as the founding program director.

Kirsten Mawyer, PhD, is an assistant professor of secondary science in the Institute for Teacher Education Secondary Program in the College of Education at the University of Hawaiʻi at Mānoa. Mawyer's teaching, program development, and research focus on how to support secondary science teachers in incorporating culturally sustaining and revitalizing pedagogy in their teaching practice in the context of UHM's commitment to serving as a Native Hawaiian place of learning. She is also the co–principal investigator on an NSF-funded project to help science educators in Hawaiʻi develop expertise in integrating knowledge of place, local and indigenous knowledge, and STEM. In addition to this work, Mawyer's research explores disciplinary literacy and designing supports for pre-service secondary science teacher learning.

Jonathan (J.D.) McCausland is a PhD student in the Department of Curriculum and Instruction within the Pennsylvania State University College of Education. He has been working on projects that examine how simulations and visualizations support students in learning about plate tectonics and risk surrounding natural hazards. In addition to this work, J.D. is involved in an ongoing research practice partnership geared toward understanding how expert and novice teachers learn Ambitious Science Teaching practices. His primary research interests focus on examining white supremacy within science and science education as well as how to prepare science teachers to enact antiracist teaching practices.

Anna MacPherson, PhD, senior manager of educational research and evaluation at the American Museum of Natural History, began her career as a high school science teacher in New York City. She received her doctorate in science education from Stanford University. Her dissertation explored how to design assessments of students' argumentation in science, for which she received the Outstanding Doctoral Research Award from the National Association for Research in Science Teaching. In her role at AMNH, she conducts research about student and teacher learning in the museum and in classrooms around the city. She is particularly interested in how students engage in the practices of constructing explanations and arguing from evidence, and how preservice and experienced teachers learn how to design and carry out ambitious science instruction. In the museum's Master of Arts in Teaching Earth Science Residency Program, she coteaches the secondary science methods course. She enjoys pressing her family for evidence-based explanations of scientific phenomena while exploring the city on the weekends and national parks during the summer.

Laura Rodriguez, PhD, is an assistant professor of science education at Eastern Connecticut State University. She teaches a variety of education courses, including elementary and secondary science methods. Her research focuses on how people develop identities as lifelong STEM learners. Specifically, she is interested in how informal intergenerational collaborations can support STEM learning and identity formation.

Déana A. Scipio, PhD, is the director of graduate and campus education programs at IslandWood, a residential environmental learning center. Scipio leverages her background in informal learning environments and design-based research to implement a multilayered, immersive learning environment at IslandWood that supports the needs of fourth- through sixth-grade students and graduate students. Her research agenda into broadening participation is grounded in justice, equity, diversity, and inclusion (JEDI). Scipio's work is focused on increasing access for participants from nondominant communities and simultaneously working to transform what it means to participate in science. This work entails antibias education with participants from dominant communities and actively transforming educational institutions. At IslandWood, Scipio is able to operationalize her broadening participation approach by working within a nonprofit organization that serves both youth and adults from nondominant and dominant communities as they learn together about the environment and sustainability. Scipio teaches courses at IslandWood in the graduate program, manages the faculty, and designs the curricular arc for the ten-month graduate residency in Education for Environment and Community. She is also studying approaches to antibias education for youth at IslandWood, pedagogical approaches to using positioning theory with adult learners, and pedagogies of joy as a guide for designing and implementing productive participant-centered learning environments.

Jessica Thompson, PhD, is an associate professor in teaching, learning, and curriculum at the University of Washington. She engages in research–practice partnerships with culturally and linguistically diverse school districts to develop and sustain Local Improvement Networks that support the improvement of ambitious and equitable science teaching. Such networks draw on the expertise of students, novice and experienced science teachers, science and emergent bilingual coaches, principals, and district leadership. She has expertise in facilitating and examining teacher learning of Ambitious Science Teaching practices at the elementary and secondary level, as well as expertise in the methods of improvement science. She also runs afterschool programs that learn from girls who are minoritized by schools, studying their engagement in scientific inquiry. Her background is in biology and chemistry, and she taught high school and middle school science as well as drop-out prevention courses in North Carolina and Washington State.

Mark Windschitl, PhD, is a professor of science education at the University of Washington. He taught secondary science for thirteen years in the Midwest before receiving his doctorate and moving to Seattle. His research focuses on how teachers take up new practices and the tools they use to engage students in authentic disciplinary activity. Windschitl is the lead author of "Rigor and Equity by Design: Seeking a Core of Practices for the Science Education Community"— a chapter in the newest edition of the American Education Research Association's *Handbook of Research on Teaching*. He is a past recipient of the AERA Presidential Award for Best Review of Research and a member of the National Research Council Committee on Strengthening and Sustaining Teachers.

Carla Zembal-Saul, PhD, is a social science researcher, science teacher educator, and biologist. She holds the Kahn Endowed Professorship in STEM Education at Pennsylvania State University. Her work is situated in school-university-community partnerships in the United States and abroad. Zembal-Saul's research investigates how preservice and practicing elementary teachers learn to engage children in productive participation in sensemaking through scientific discourse and practices. Her most recent work is situated in a community undergoing rapid demographic shifts with teachers and other school professionals who work with emergent bilingual students and their families. Zembal-Saul is committed to collaborating with teachers and families, bridging research and practice, and codesigning and coauthoring with practitioners. She was recognized as a National Science Teachers Association Fellow in 2015, and she served on the National Academies of Sciences, Engineering, and Medicine's Board on Science

Education consensus committee that wrote the report *Science Teachers' Learning: Enhancing Opportunities, Creating Supporting Contexts* (2015).

Yang Zhang is a third-year PhD student in the Department of Teaching and Curriculum within Warner School of Education at the University of Rochester. She has been working on different research projects, including the impact of augmented reality on chemical engineering majors' learning at the higher education level, consequential learning, and culturally sustaining Ambitious Science Teaching. Her primary research interest focuses on K–12 science teachers' professional identity development with an orientation toward equity.

About the Editors

David Stroupe, PhD, is an associate professor of teacher education and science education at Michigan State University. He also serves as the associate director of STEM Teacher Education at the CREATE for STEM Institute at MSU. He has three overlapping areas of research interests anchored around ambitious teaching practice. First, he frames classrooms as science practice communities. Using lenses from Science, Technology, and Society and the History and Philosophy of Science, he examines how teachers and students negotiate power, knowledge, and epistemic agency. Second, he examines how beginning teachers learn from practice in and across their varied contexts. Third, he studies how teacher preparation programs can provide support and opportunities for beginning teachers to learn from practice. Stroupe is the recipient of the AERA Exemplary Research Award for Division K (Teaching and Teacher Education), and the Early Career Research Award from the National Association for Research in Science Teaching. He also has a background in biology and taught secondary life science for four years.

Karen Hammerness, PhD, is the director of educational research and evaluation at the American Museum of Natural History. At the museum, her research centers on educator and youth learning inside and outside of school, and how equity and access can be addressed. She is especially interested in the professional development of science teachers along the teacher learning continuum (from preservice to novice to experienced) and in how teacher educators support teachers' exploration of theoretical principles, vision, and classroom practices. She is currently a co–principal investigator of two research studies funded by the National Science Foundation: one follows the trajectories of New York City youth who have been mentored in an immersive science research program (across twenty-one informal science institutions); the other examines the development of a Next Generation Science Standards middle school ecology curriculum and professional development experiences for teachers. She also researches the design and pedagogy of teacher education not only in the United States, but also internationally, examining a variety of teacher education programs from teacher residencies to "context-specific" programs to college and university-based programs both in the United States and in other countries, including Chile, Finland, and Norway. Hammerness is a coeditor of *Inspiring Teaching: Preparing Teachers to Succeed in Mission-Driven Schools* (Harvard Education Press) and the lead author of *Empowered Educators in Finland: How Leading Nations Design Systems for Teaching Quality* (Jossey-Bass).

Scott McDonald, PhD, is a professor of science education at the Pennsylvania State University and director of the Krause Studios for Innovation in the PSU College of Education. He received his undergraduate degree in physics with a focus on astronomy and astrophysics. He was a high school physics teacher for six years before returning for a PhD in learning technologies and science education at the University of Michigan. McDonald takes a design-based approach to research on science teacher learning and student learning in the geosciences. His research focuses specifically on teacher learning, framed as professional pedagogical vision for ambitious and equitable science teaching practices. He is also engaged in the development of learning progressions in earth and space sciences as part of a series of NSF-funded projects, including Geological Models for Explorations of Dynamic Earth (GEODE) and GeoHazard: Modeling Natural Hazards and Assessing Risks.

Acknowledgments

This book is the culmination of conversations, conferences, and collaborations that have been occurring for years. In particular, we thank John Settlage and Adam Johnston for providing the initial Crossroads space for vexations and ventures around core practices and ambitious instruction, and for the encouragement to engage in conversations across programs. We are extremely appreciative to the National Science Foundation and Michael Ford for providing us with funds to meet as a group of colleagues, from which this book emerged. We also thank colleagues in the Michigan State University College of Education for their assistance with the meeting of science teacher educator colleagues. In particular, the CREATE for STEM Institute (Joe Krajcik, Bob Geier, Ligita Espinosa, Sue Carpenter, Alison Vincent, and Dez Thomas) and the Department of Teacher Education (Margaret Crocco, Terry Edwards, Kristi Lowrie, and Sue Sipkovsky) provided much-needed support. We are also thankful for the wonderful work of the graduate students at Michigan State University who helped during the meeting. Led by Sinead Brien, the graduate student volunteers included Julie Christensen, Krista Damery, Brian Hancock, Christa Haverly, May Lee, and Kraig Wray. We also had the privilege of hearing from teachers who participated in a practice-based teacher preparation program, and we are thankful to Amanda McSween, Caeli Loris, Ryan Hibbs, Katelyn O'Brien, Claire Morrison, and Kathryn Schwartz for sharing their time and insights as colleagues who engage with core practices daily in classrooms. We are also deeply indebted to Ramya Swayamprakash for her careful edits and guidance on formatting this book. We are thankful for conversations with colleagues who push our thinking as we all strive for equitable teaching and learning in classrooms. Specifically, we thank Thomas Philip, Ilana Horn, Manka Varghese, Mariana Souto-Manning, Dorinda Carter Andrews, Jamy Stillman, Julie Contino, Robert Steiner, Natasha Cooke-Nieves, and David Hammer for discussing and reading sections of the book. Such conversations helped us clarify ideas, engage with complex challenges, and truly listen to important perspectives on the work. Finally, we are very thankful for the team at Harvard Education Press, including Caroline Chauncey for her wonderful guidance and amazing insights. We are also thankful to Pam Grossman for editing the series about core practices, and for encouraging this book's progress.

On a personal level, we are grateful for our families, who encourage and support us throughout our work. David wishes to thank Erin, Emma, and Zoe for their encouragement, love, and laughter. Karen is grateful for the listening ear and constant support of Thomas and their three daughters, Hannah, Clara, and Stella, and excited that they enjoy science and might someday work with teachers some of us have prepared. Scott wants to thank Grace and Emily for their love and support, and Christine for being a wonderful partner and coinquirer in teacher education.

Index